8/21

1

Florida's Butterflies
and Other Insects

Palamedes swallowtail

Pterourus palamedes

Florida's Butterflies
and Other Insects

by
Peter D. Stiling

Pineapple Press  Sarasota, Florida

For Joan and John

Published by Pineapple Press, Inc., P.O. Drawer 16008, Sarasota, Florida 34239.

Library of Congress Cataloging-in-Publication Data

Stiling. Peter D.
    Florida's butterflies and other insects / by Peter D. Stiling.—
1st ed.
            p.   cm.
    Bibliography: p.
    Includes index.
    ISBN 0-910923-54-X : $24.95
    1. Butterflies—Florida—Identification.   2. Insects—Florida—
Identification,   I. Title.
QL551.F6S75 1988
595.78'09759—dc19   88-25061
                                                            CIP

First Edition
            10 9 8 7 6 5 4 3 2 1

Printed in Singapore
    through Palace Press
Typography by Hillsboro Printing
    Tampa, Florida
Design by Joan Lange Kresek

$\mathcal{T}$able of contents

Acknowledgments

I should like to thank Michael Antolin, Brent Brodbeck, John Lyons, Fay Phillips, Colin Phipps, and Sharon Strauss for collecting insects for me or for bringing them to my attention. In this vein, Steve Atkins and Joe Bigelow were constant sources of inspiration and of amazing bugs. Nick Livingstone deserves special mention for his design and construction of camera accessories, gratis, in particular for a pair of flash brackets. I should also like to acknowledge here the debt I owe to Anne Thistle for the careful editing and typing of this manuscript and many others. Thanks also to the staff of the insect collection of the Florida Department of Agriculture and Consumer Services, Division of Plant Industry, at Gainesville for some species identifications. Dr. R. M. Baronowski, Tropical Research and Education Center, Homestead; Dr. John B. Heppner, Division of Plant Industry, Gainesville; and Drs. S. Dunkel, G. B. Edwards, G. B. Fairchild, D. H. Habeck, F. Mead, L. Stange, T. Walker, and R. Woodruff, Department of Entomology, Gainesville, provided useful comments on the manuscript. And finally a big thank-you to Jacqui for accompanying me on hot walks in the bush and for putting up with caterpillar frass in the kitchen. Further pertinent information on Florida's insects would be much appreciated by the author.

List of plates

1.	Giant swallowtail	*Heraclides cresphontes*
2.	Pipevine swallowtail	*Battus philenor*
3.	Gold rim	*Battus polydamus*
4.	Eastern black swallowtail	*Papilio polyxenes*
5.	Tiger swallowtail	*Pterourus glaucus*, yellow form
6.	Tiger swallowtail	*Pterourus glaucus*, black form
7.	Palamedes swallowtail	*Pterourus palamedes*
8.	Green-clouded swallowtail	*Pterourus troilus*
9.	Zebra swallowtail	*Eurytides marcellus*
10.	Schaus' swallowtail	*Heraclides aristodemus ponceanus*
11.	Zebra longwing	*Heliconius charitonius*
12.	Gulf fritillary	*Agraulis vanillae*
13.	Gulf fritillary, aberration fumosus	*Agraulis vanillae nigrior* aberration *fumosus*
14.	Julia or flambeau	*Dryas iulia*
15.	Buckeye	*Junonia coenia*
16.	Caribbean buckeye	*Junonia evarete*
17.	White peacock	*Anartia jatrophae*
18.	Malachite	*Siproeta stelenes*
19.	Red-spotted purple	*Basilarchia astyanax*
20.	Viceroy	*Basilarchia archippus*
21.	Florida purplewing	*Eunica tatila*
22.	Ruddy daggerwing	*Marpesia petreus*
23.	Variegated fritillary	*Euptoieta claudia*
24.	Pearl crescent	*Phycoides tharos*
25.	Question mark	*Polygonia interrogationis*
26.	Painted lady	*Vanessa cardui*
27.	Florida white	*Glutophrissa drusilla*
28.	Great southern white	*Ascia monuste*
29.	Sleepy orange	*Eurema niccipe*
30.	Cloudless sulphur	*Phoebis sennae*
31.	Orange-barred sulphur	*Phoebis philea*
32.	Silver-spotted skipper	*Epargyreus clarus*
33.	Long-tailed skipper	*Urbanus proteus*
34.	Tropical checkered skipper	*Pyrgus oileus*
35.	Atala	*Eumaeus atala*
36.	Red-banded hairstreak	*Calycopis cercops*
37.	Miami blue	*Hemiargus thomasi*
38.	Large wood nymph	*Cercyonis pegala*
39.	Monarch	*Danaus plexippus*
40.	Queen	*Danaus gilippus*
41.	Tulip-tree silkmoth	*Callosamia angulifera*
42.	Luna moth	*Actias luna*
43.	Io moth	*Automeris io*
44.	Polyphemus moth	*Antheraea polyphemus*
45.	Spiny oakworm	*Anisota stigma*
46.	Rosy maple moth	*Dryocampa rubicunda*
47.	Regal moth	*Citheronia regalis*
48.	Imperial moth	*Eacles imperialis*
49.	Pink-spotted hawk moth	*Agrius cingulatus*
50.	Rustic sphinx	*Manduca rustica*
51.	Catalpa sphinx	*Ceratomia catalpae*
52.	Twin-spotted sphinx	*Smerinthus jamaicensis*

7

53. Blinded sphinx	*Paonias excaecatus*
54. Small-eyed sphinx	*Paonias myops*
55. Banded sphinx	*Eumorpha fasciata*
56. Hydrangea sphinx	*Darapsa versicolor*
57. Hog sphinx	*Darapsa myron*
58. Tersa sphinx	*Xylophanes tersa*
59. Rattlebox moth	*Utethesia bella*
60. Giant leopard moth	*Ecpantheria scribonia*
61. Spanish moth	*Xanthopastis timais*
62. Underwing	*Catocala sp.*
63. Eastern black swallowtail caterpillar	*Papilio polyxenes*
64. Orange dog caterpillar	*Heraclides cresphontes*
65. Brazilian skipper caterpillar	*Calpodes ethlius*
66. Gulf fritillary caterpillar	*Agraulis vanillae*
67. Polyphemus caterpillar	*Antheraea polyphemus*
68. Luna moth caterpillar	*Actias luna*
69. Hickory horned devil	*Citheronia regalis*
70. Tobacco hornworm	*Manduca sexta*
71. Eastern tent caterpillar	*Malacosoma americanum*
72. Bagworm	*Thyridopteryx ephemeraeformis*
73. White-marked tussock moth caterpillar	*Orgyia leucostigma*
74. Pale tussock moth caterpillar	*Halysidota tessellaris*
75. Yellow-necked caterpillar	*Datana ministra*
76. Red-humped caterpillar	*Schizura concinna*
77. Spiny oakworm	*Anisota stigma*
78. Saddleback caterpillar	*Sibine stimulea*
79. Hag moth caterpillar	*Phobetron pithecium*
80. Brown-spotted yellow-wing	*Celithemis eponina*
81. Green clearwing	*Erythemis simplicicollis*
82. Swift long-winged skimmer	*Pachydiplax longipennis*
83. Comet darner	*Anax longipes*
84. Doubleday's bluet	*Enallagma doubledayii*
85. Black-winged damselfly	*Calopteryx maculata*
86. Variable dancer	*Argia fumipennis*
87. Southeastern lubber grasshopper	*Romalea microptera*
88. Lubber grasshopper nymph	*Romalea microptera*
89. Bird grasshopper	*Schistocerca alutacea*
90. American bird grasshopper	*Schistocerca americana*
91. Carolina locust	*Schistocerca damnifica*
92. Angular-winged katydid	*Microcentrum retinerve*
93. Field cricket	*Gryllus sp.*
94. Mole cricket	*Scapteriscus sp.*
95. St. Andrew's cotton bug	*Dysdercus andreae*
96. Big-legged bug	*Acanthocephala declivis*
97. Largid bug	*Largus succinctus*
98. Wheel bug	*Arilus cristatus*
99. Southern green stink bug	*Nezara viridula*
100. Brochymena bug	*Brochymena arborea*
101. Giant water bug	*Lethocerus griseus*
102. Cicada	*Tibicen sp.*
103. Treehopper	*Entylia carinata*
104. Glassy-winged sharpshooter	*Homalodisca coagulata*

105. Black horsefly	*Tabanus atratus*
106. Tachinid fly	*Archytas* sp.
107. Love bug	*Plecia nearctica*
108. Bee hunter	*Laphria* sp.
109. Long-legged fly	*Condylostylus* sp.
110. Paper wasp	*Polistes* sp
111. Potter wasp nest	*Eumenes fraternus*
112. Thread-waisted wasp	*Eremnophila aureonotata*
113. Mud dauber nest	
114. Hunting wasp	*Campsomeris quadrimaculata*
115. Honey bee	*Apis mellifera*
116. American bumble bee	*Bombus pennsylvanicus*
117. Halticid bee	*Augochlora* sp.
118. Fire ant	*Solenopsis invicta*
119. Cow killer	*Dasymutilla occidentalis*
120. Patent leather beetle	*Odontotaenius disjunctus*
121. Unicorn beetle	*Dynastes tityus*
122. Ox beetle	*Strategus antaeus*
123. Larva of ox beetle	*Strategus antaeus*
124. Green June beetle	*Cotinus nitida*
125. Spotted pelidnota	*Pelidnota punctata*
126. Palm weevil	*Rhynchophorus cruentatus*
127. Tortoise beetle	*Hemisphaerota cyanea*
128. Carolina sawyer	*Monochamus carolinensis*
129. Ivory-spotted borer	*Eburia quadrigeminata*
130. Cylindrical hardwood borer	*Neoclytus acuminatus*
131. Blind click beetle	*Alaus myops*
132. Tiger beetle	*Cicindela scutellaris*
133. Fiery searcher	*Calosoma scrutator*
134. Firefly	*Photinus* sp.
135. Palmetto stick insect	*Anisomorpha buprestoides*
136. Praying mantis	*Stagmomantis carolina*
137. American cockroach	*Periplaneta americana*
138. Gall produced by gall insect	
139. Leaf mine produced by leaf miner	*Tischeria citripennella*
140. Hentz's striped scorpion	*Centruroides hentzi*
141. Daddy-long-legs	*Leiobunum* sp.
142. Tick	*Dermacentor* sp.
143. Golden silk spider	*Nephila clavipes*
144. Black and yellow argiope, adult	*Argiope aurantia*
145. Black and yellow argiope, juvenile	*Argiope aurantia*
146. Silver argiope	*Argiope argentata*
147. Arrow-shaped micrathena	*Micrathena sagittata*
148. Crablike spiny orb weaver	*Gasteracantha cancriformis*
149. Orchard spider	*Leucauge venusta*
150. Green lynx spider	*Peucetia viridans*
151. Carolina wolf spider	*Lycosa carolinensis*
152. Cyclosa spider	*Cyclosa turbinata*
153. Millipede	*Narceus* sp.
154. Centipede	*Scolopendra* sp.

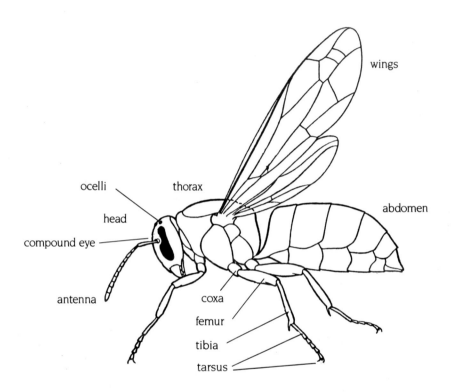

A typical insect showing the major features of the insect body

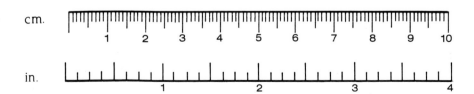

Introduction

For every living person on earth there are estimated to be 200 million insects. This number equates to roughly 10 billion in each square kilometer of habitable land or 26 billion per square mile. Not only are insects astonishingly common, they are a very diverse group, with 850,000 described species—roughly 85% of the known living things on the planet and far more than any other group such as the fish, birds, or mammals. Even in cool temperate England, where insects are not so common as in Florida, there are at least 10 species of insect to each plant species on which they feed. Chances are then that both the resident Floridian and the visitor to the Sunshine State will meet up with many insects during their tenure in Florida. This book is designed to help identify some of the more common and visible species of Florida. It would be impossible to identify them all in such a small volume, so only the more interesting, colorful, or often-encountered types are treated.

Florida's southerly location between 24° N and 31° N, together with the action of the Gulf Stream, gives it a subtropical climate and flora unlike any other state. Pine flatwoods are the predominant vegetation community and consist of long-leaf, slash, or pond pine mixed with oaks. Herbs, saw palmetto, small hardwood trees, and other shrubs form an understory. Swamp forests, saltwater marshes, and, south of Tampa, mangroves rim the peninsula. A large number of Florida's insect fauna that feed on this vegetation are truly tropical in origin, having colonized the area from the West Indies and the Yucatan peninsula. Despite this tropical connection Florida is depauperate in terms of number of species as compared to an equivalent area of Central America. There are no high mountain ranges in Florida to provide an array of habitats, but perhaps more important is the recent geologic past, in particular the four ice ages that have occurred in the last million years. Although ice sheets never reached Florida itself, temperatures dropped by 5–10° and precipitation patterns changed. South Florida in this period probably had a fauna much like that of northern Georgia and the Carolinas today and had no West Indian species. Perhaps even more importantly, a large volume of water was locked up as ice, so the sea levels fell as much as 300 feet, dramatically increasing the size of Florida. In the warm interglacials, however, the ice melted and the sea levels rose to cover most of what constitutes today's peninsula. Each time this cycle occurred, much of the fauna and flora was gradually washed away or squeezed out from the few remaining islets. The present-day fauna is evolutionarily very young, having developed only in the last 2,000–3,000 years. During that time there have been relatively few West Indian colonists, together with some from the northern continental United States.

The body designs and life cycles of insects have, however, helped them colonize Florida more rapidly than some other animals. To begin with, most species fly, and some, such as the monarch butterfly, actively undertake long migrations from overwintering grounds in Mexico to the United States. Some have even been known to cross the Atlantic to Europe. Second, insects are small. They do not actively "breathe," most having no specialized lungs. Instead, oxygen diffuses through pores in the "skin" (actually a chitinous exoskeleton) along tubelike tracheae to cells on the inside. Diffusion is a relatively slow and inefficient method of respiration, and gases cannot diffuse very far into tissue. Dog-sized insects would suffocate very quickly. This method of respiration restricts species to small, or at least thin, bodies. Small body size, however, is concommitant with short generation time, and fast generations are more quickly responsive to evolutionary pressures, with the result that one species can rapidly diversify to many, each often colonizing a different microhabitat. Here again, small body size is advantageous because insects can live under bark, in soil, within leaves, under rocks, and so on, places where larger animals would never fit. Small body size also comes with the advantage of a smaller mass, more easily protected by the tough exoskeleton. Many species, then, are robust and able to withstand forces that one might think would wreck them. Finally, 88% of insects undergo a complete metamorphosis; that is, the adult develops from a form such as a caterpillar, grub, or maggot that bears no resemblance to it. This life history ensures that no habitat goes uncolonized; although some patches like dung or stagnant water would ruin an adult's wings, the larval insect can exploit the resource with ease. Only in a few groups, such as the Hemiptera, Homoptera, and Orthoptera, do "baby" insects (called nymphs) look like miniatures of the adults; they also feed on the same thing—plants.

Besides the beauty of many of Florida's butterflies and other insects, it is worth remembering that many insects serve very positive purposes. The fabric of society would be dramatically changed without them—quite literally in the case of the imported silk moth, *Bombyx mori*. Bees alone provide a pollination service to plants estimated to be worth at least $4.5 billion annually in the United States. Beetles and flies dispose of dead vegetation, animal corpses, and dung, returning vast amounts of nutrients to the soil.

Of course, many insects are serious agricultural pests. Damage and control cost Florida $1.3 billion a year, about $125 for each Floridian (Meeker 1987). The threat of medfly infestation spreads panic among citrus growers in the state, as grub-ridden fruit is offensive to the buying public. In a worst-case scenario, medflies would totally destroy the fruit. Other serious pests of oranges include scale insects, which look like miniature white cotton balls on leaves. The "orange dog" butterfly larva, which chews the foliage (see page 46), is a minor citrus pest. Many other insects attack different crops. Armyworms, earworms, cabbage loopers, and cutworms feed on vegetables, corn, crucifers, and a host of other plants. Ornamentals too have their insect enemies. For example, poinsettias are attacked by hornworms, and bougainvillea by pyralid moth larvae. A whole arsenal of sprays is marshalled against these pests, but so far none of them has been eliminated. A classic reference to the abundance of these life forms in Florida was made by the first federal entomologist, Townsend Glover, in his "Florida Litany":

From red bugs and bed bugs,
from sand flies and land flies,
 Mosquitoes, gallinippers and fleas,
 From hog ticks and dog ticks
and hen lice and men lice,
 We pray thee, good Lord, give us ease.

All the congregation shall scratch
and say Amen.

Insects can act as vectors for the spread of diseases, and we should be thankful that such diseases as malaria and yellow fever are now largely absent from Florida. At one time yellow fever was rampant in the state; an epidemic at St. Josephs (now Port St. Joe) in 1848 destroyed 75% of the population, and the town was abandoned shortly thereafter. The city of Jacksonville was noted for a great pestilence in 1857, brought on by a malignant strain of yellow fever. Steamers would not stop, and Jacksonville became isolated from the rest of the world. In all, 127 people died (Zak 1986). In 1888, an even worse epidemic broke out, and by Thanksgiving 5,000 people had succumbed to yellow fever, with 400 dead. At that time, of course, people were pitifully ignorant of the causes of such diseases. In the early stages of the 1888 epidemic in Jacksonville, a battery of six cannons was fired to combat the disease. The theory was that explosion of gunpowder at night would destroy the germs of the disease by concussion of the atmosphere. After five nights, the ammunition was exhausted, and the only clear result was that the patients could not endure the noise (Madden 1945).

As for malaria, the peak year of reported cases was 1919, when 1,895 cases were found (Lieux 1951). In 1929, a record 470 people died from the disease, and this fully 30 years after malaria was shown to be carried by mosquitoes in 1897. In the early years only programs designed to educate the public were carried out. The first substantial malaria-control project in Florida was conducted between 1920 and 1921 at Perry, where 65% of the population were afflicted. Even as late as 1939, the death rate from malaria was 17.3 per 100,000 people, third highest in the U.S. It is ironic to note that the struggle against malaria inadvertently led to air conditioning. Dr. John Gorrie at Apalachicola, in his attempts to cure malaria, first invented a machine to cool patients down—an invention that eventually led to the air conditioner (Madden 1945).

Of course the latest scare was that mosquitoes could transmit AIDS and that Floridians, with such a high number of mosquitoes in the state, would be in serious trouble. Concern about a possible role for insects in transmitting HIV (human immunodeficiency virus) arose in part because of the high frequency of AIDS in swampy Belle Glade. Over the past five years, 76 people out of the population of 16,500 contracted the disease—a rate of 461 per 100,000, comparable to that in the areas of highest risk, San Francisco and New York City. Scientists confirmed that mosquitoes can retain the virus in their digestive systems for a short time but showed that it cannot replicate there. Experts have repeated that there is no evidence and scant likelihood that mosquitoes could transfer AIDS to other organisms. Investigators from the Centers for Disease Control found that the use of intravenous drugs, prostitution, and promiscuity are common in Belle Glade and that many of the people there come from Haiti, where there is a high level of infection with HIV. These were the likely reasons behind the high incidence of AIDS in that city (Kingman 1987).

It is important to remember that we are much better off with insects than without them. Experiments involving cages over orchard trees have shown that, if bees are excluded, the set of fruit is less than 1% of the blooms, whereas with the bees present it is up to 44%. Furthermore, introducing insects to feed on alien weeds that have run rampant in the state, such as Brazilian pepper, *Schinus terebinthifolius*, represents probably the only cost-effective method of controlling these plants. Some insects destroy the seeds, preventing the weed's spread, and others chew the foliage, reducing the extent of existing infestations. In the 1950s and 1960s, many of Florida's inland waterways were choked with an exotic South American weed, alligatorweed. In 1965 a pretty leaf-eating beetle was introduced into Florida from Argentina to control it. Only 250 beetles were originally released, at Jacksonville, but these multiplied rapidly and spread. Control of alligatorweed was achieved—the world's

first successful introduction of an insect against an aquatic weed. Hopes are high that introduced weevils will be able to control water hyacinth, another plague of Florida's waterways (Buckingham 1987).

In this book, common names are used for insects whenever possible, together with their scientific designations (always in italics). The latter are two-part Latin names, each unique to a particular type of insect. The first part of this name designates a genus, and the second part is a species name. A genus is a group of closely related species; for example, the genus *Battus* contains only some closely related swallowtail butterflies. Genera of similar characteristics are combined to form families. All the swallowtails are united in the family Papilionidae. A group of similar families forms an order, like Lepidoptera (the moths and butterflies) or Coleoptera (the beetles), and similar orders form a class. Some common classes include Insecta (the insects), Aves (the birds), and Mammalia (the mammals). This scientific system of nomenclature was established by a Swedish naturalist, Carl von Linné, in 1758 and has been used ever since. With so many different kinds of insects, often superficially quite similar, we would soon be mired in a tangle of duplicated and overlapping common names, so the scientific system is necessary to ensure that each species has its own characteristic name. Even for groups such as the moths and butterflies, in which common names often correspond quite well to individual species, Latin names ensure that students of entomology all around the world will know what species is meant by "*Danaus plexipus*," even if only English speakers know it as the monarch butterfly. As Alice in Wonderland said to her friend the gnat, a name is "no use to them, but it's useful to the people that name them."

Florida's Butterflies
and Other Insects

Butterflies and moths: order LEPIDOPTERA

Of all life on earth, butterflies evoke some of the strongest feelings, most often admiration of their great beauty and grace. They have been, and still are, the subjects of paintings and poems, of dress designs and party costumes. The order Lepidoptera, to which butterflies and moths belong, includes about 11,230 North American species, but only 760 of these are butterflies. The rest are moths, the majority of which are very small and have dull colorations but a few of which are as colorful as butterflies. Butterflies on the other hand are often dazzlingly colored. This coloration is the result of two mechanisms that operate independently or in combination. Lepidoptera literally means scale wing, and the wing scales can be hollow, with a smooth lower surface and a ridged upper surface. This structure diffracts light, producing iridescent sheens and colors. Colors can also be laid down inside the scales, producing pigmentary colors. Butterflies are usually diurnal and have clubbed antennae. Moths are most commonly nocturnal, have pectinate (branched) antennae (at least for the larger species), and possess a frenulum, a hooklike mechanism that couples front and hind wings together so that they beat as one in flight. In butterflies, front and hind wings lack a frenulum but overlap greatly to enable the wings to synchronize strokes.

Florida has about 2600 species of Lepidoptera. South Florida alone, from Tampa to Vero Beach southward, has 119 species of butterflies, the majority (50 species) being dull-colored skippers, members of the hesperiid family (Kimball 1965, Scott 1972). There are 27 nymphalids, 17 pierids or sulphurs, and 9 papilionids or swallowtails. Of the south Floridian butterflies, 69 are most closely related to those of the coastal plain of the southeastern United States, 20 are related to species from the Antilles (9 of which are found only in southern Florida and the Antilles), two are most closely related to Mexican species, and 25 are equally closely related to Antillean and Mexican subspecies of the species. Even though Florida has a continental connection to Central America, via Mexico and the Gulf Coast states, it seems most tropical species have arrived from the Antillean fauna. The Florida Keys are certainly biogeographically closely related to the Antilles, and Key West is closer to Havana than to Miami. Of the 80 or so species of butterfly in the Keys, 50 are also found in Cuba, and it is probably true that migration across water has caused the faunas of these two regions to be fairly similar: the index of faunal resemblance between the Keys and Cuba is at least 73% (Scott 1972).

1. Giant swallowtail
Heraclides cresphontes

2. Pipevine swallowtail
Battus philenor

3. Gold rim
Battus polydamus

4. Eastern black swallowtail
Papilio polyxenes

5. Tiger swallowtail
Pterourus glaucus, yellow form

6. Tiger swallowtail
   Pterourus glaucus, black form

7. Palamedes swallowtail
   Pterourus palamedes

8. Green-clouded swallowtail
   Pterourus troilus

9. Zebra swallowtail
   Eurytides marcellus

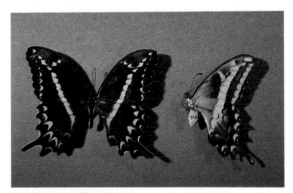

10. Schaus' swallowtail
    Heraclides aristodemus ponceanus

Butterflies and Moths: Order Lepidoptera ～ 19

Swallowtails: Papilionidae

Swallowtails, named for the long tails that project from the hind wings, are among Florida's most striking butterflies and can be frequently seen at flowers. So typical of butterflies are members of this group that Linnaeus named them using the Latin word for butterfly, *papilio*. The family contains about 700 species distributed throughout the world, with about 30 North American species and 8 common in Florida, plus the rarely seen Schaus' swallowtail, *Heraclides aristodemus*. In addition, a tenth swallowtail, Androgeus' swallowtail, *Papilio androgeus*, was recorded in Broward County in the 1970s, and it has been known to breed on oranges there (Leston and Waddill 1980). This swallowtail is normally resident only in South America and the Greater Antilles.

Papilionids can be broadly divided into three groups, true swallowtails (*Papilio*); kite swallowtails, which have narrow pointed wings, giving the appearance of an old-fashioned kite (*Eurytides*); and poison eaters (*Battus*), so called because the larvae feed on *Aristolochia* vines, from which they derive poisonous substances. All the larvae are smooth skinned and possess a strange structure called an osmeterium, a forked horn that can be erected from behind the caterpillar's head. It emits a foul-smelling odor and is particularly useful in detering would-be predators and especially parasites. The adult butterflies are very colorful, with black and yellow predominating. *Papilio* sp. are generally black with yellow spots or bands, giant swallowtails (*Heraclides* sp.) brown and yellow, tiger swallowtails yellow with black stripes, and pipevine swallowtails (*Battus* sp.) blackish and blue-green. The bright colors generally advertise distastefulness to predatory birds.

Giant swallowtail: *Heraclides cresphontes*   Fig. 1

The giant swallowtail is America's largest butterfly, with a wingspan of up to 15 cm, though in Florida it is rivalled by the tiger swallowtail. It is a widespread and strong-flying species ranging from southern Canada across most of the eastern United States and down into Mexico. The larva, called the "orange dog" by orange growers, is rarely considered a pest on citrus and then only on young trees (see Fig. 64).

Pipevine swallowtail: *Battus philenor*   Fig. 2

The wings of this butterfly are usually black with a blue-green iridescence, which gives rise to its alternative name, "blue swallowtail." A wide-ranging species, it occurs throughout the United States except the extreme north.

Horticulture has caused the spread of pipevines, *Aristolochia* sp., the larval food of this butterfly, thereby extending its range. The pipevines are noxious, giving the larvae and hence the adults a bad taste. As a result, many birds learn to reject pipevine swallowtails as food.

Other species such as the spicebush swallowtail and red-spotted purple (a nymphalid) are similar in color and may have evolved to mimic B. *philenor* (Brower and Brower 1962). Such similarity, where mimics are also protected from predators by a warning coloration, is known as Batesian mimicry (see page 00).

Gold rim: *Battus polydamus*   Fig. 3

An easily recognized swallowtail, the gold rim is tailless and has a band of yellow spots within the margins of the front and hind wings. It is locally distributed in Florida, with a general range south of a line from Cross City to Gainesville and Palatka. Its flight period is generally from April through November. Like those of the pipevine swallowtail, larvae feed on *Aristolochia* vines, which makes adults distasteful to birds.

Eastern black swallowtail: *Papilio polyxenes* — Fig. 4

This common swallowtail often flies low to the ground, drifting and stalling erratically before coming to rest on low vegetation. Common from March to May and measuring about 6.5–9 centimeters in wingspan, this is the smallest of the predominantly black swallowtails with yellow markings. Caterpillars are white to leaf green, with black bands on each segment (see Fig. 63).

Tiger swallowtail: *Pterourus glaucus* — Fig. 5 & 6

The tiger swallowtail exhibits a striking form of dimorphism, which is confined to the female sex. Males and some females are yellow with black tiger stripes across the wings. Some females, however, exhibit a dark form, being black above with a bordering of yellow, blue, and orange. This form is generally predominant south of latitude 40°N and is thought to be a mimic of the pipevine swallowtail. The larva of the tiger swallowtail resembles a bird dropping when small but when mature is green and swollen in front and has large false eyespots. It can be found on a wide variety of broadleaf trees and shrubs, but in south Florida it has only been recorded on sweetbay, *Magnolia virginiana* (Scriber 1986). The earliest known picture of an American butterfly is one of the tiger swallowtail painted by John White, commander of Sir Walter Raleigh's third expedition to "Virginia" in 1587. In Florida, tiger swallowtails can be very large, even exceeding the giant swallowtail's size.

Palamedes swallowtail: *Pterourus palamedes* — Fig. 7

This butterfly is often more commonly encountered in coastal areas, where its habitat of subtropical wetland or humid woods with standing water may be present. It is especially common also in big swamps such as the Everglades, Big Cypress, and Okefenokee. The larva is grass green with eyespots containing a black "pupil" in an "iris" of orange.

Green-clouded or spicebush swallowtail: *Pterourus troilus* — Fig. 8

Although it is named for the distinct blue-green wash on the hindwings, this butterfly's wings also have a series of pale yellow crescentlike marks (lunules) along the margins. Caterpillars curl the leaves on which they feed and hide, commonly those of spicebush, sassafras, and sweet bay.

Zebra swallowtail: *Eurytides marcellus* — Fig. 9

The zebra swallowtail is a distinctive butterfly with long swordlike tails issuing from the hindwing. It is patterned with black stripes on a white background. It has at least three generations per year in the South, early adults being smaller and paler. Of all the swallowtails in Florida, I have found this one to be more restricted to deeper woodland than the others, which prefer open and disturbed habitats. Its range is regulated by supply of its only larval food, wild pawpaw and its relatives, which are found more commonly in pine woodlands.

Schaus' swallowtail: *Heraclides aristodemus ponceanus* — Fig. 10

Taxonomically, *Heraclides aristodemus ponceanus* is a subspecies of *H. aristodemus*, the so-called dusky swallowtail of the Greater Antilles (Stiling 1987b). Schaus' swallowtail is found only in Florida and has been listed as critically imperilled globally. Once common in the Miami area, it has long since been excluded from that area by development (Loftus and Kushlan 1984) and was only recently rediscovered on the Biscayne National Monument by Brown and his co-workers (Brown 1973). However, plantings of its natural host, torchwood, may encourage an expansion of its range, which is currently confined mainly to Elliott Key and north Key Largo (Morrison 1981).

Heliconias: Heliconiidae

The family Heliconiidae comprises just under 70 neotropical species, four of which are resident in the southern United States, three in Florida. They are characterized by a narrow wing shape, long antennae, a thin elongated abdomen, and remarkably constant wing length of between 60 and 100 mm. Some authorities regard heliconias as a separate subset of the Nymphalidae. Most have brilliant colors and distinctive patterns, a warning sign to predators of nauseating body fluids. The spinous larvae feed exclusively on *Passiflora* species, the passion flowers. Heliconias are fairly easy to keep in captivity and are long-lived. The naturalist William Beebe kept a pet *Heliconius charitonius* called Higgins for several months.

Zebra longwing: *Heliconius charitonius* Fig. 11

The zebra longwing is completely distinctive, with long black wings banded with lemon yellow. It has slow wafting flight and a tendency to roost communally at night (Jones 1930, Mallet 1986). Hammocks and thickets in Everglades National Park are good places to see these butterflies. They are not common in the northern half of the state but may reach Gainesville in the fall. The caterpillars feed on the passion flower, *Passiflora* sp.

The communal roosts contain many individuals, each of which returns faithfully each night to the same perch to sleep. So sound is their sleep that one can pick a butterfly off its roost and return it later, without waking any of the others.

Gulf fritillary: *Agraulis vanillae* Fig. 12 & 13

Another instantly recognizable species, the Gulf fritillary is brilliant red-orange above with silver-white teardrops underneath. As its name implies, this species haunts the Gulf of Mexico area and may sometimes be seen over water. It ranges into more northern states but is limited by cold weather, which neither it nor its *Passiflora* host can withstand. In the fall, Gulf fritillaries migrate southward, flying at a height of three to six feet and at about 10 mph, although this speed depends greatly on the wind at the time. On encountering a building or other obstacle, migrating fritillaries, like other butterflies, fly up and over it without changing direction (Arbogast 1966). Melanic specimens, their wings suffused to varying degrees with black, are taken from time to time (Fig. 13 and Harris 1972). The Gulf fritillary is commonly found at any of at least seven plant species during the year (May 1988) but on *Richardia scabra*, *Verbena brasiliensis*, and *Bidens pilosa* most commonly.

Julia or flambeau: *Dryas iulia* Fig. 14

Another bright orange butterfly, the Julia has no silver on the underside. Restricted to Dade and Monroe counties, it can sometimes become locally abundant in the Florida Keys. Present in hammocks and gardens, this is the swiftest flier among the United States heliconias.

11. Zebra longwing
    Heliconius charitonius

12. Gulf fritillary
    Agraulis vanillae

13. Gulf fritillary, aberration fumosus
    Agraulis vanillae nigrior aberration *fumosus*

14. Julia or flambeau
    Dryas iulia

The Nymphalidae are often referred to as the brush-footed butterflies, because their unifying characteristic is a pair of vestigial forelegs, useless for walking but often dense with scales. There are few other common features of the family; wing shape, size, and color come in a staggering array.

There are over 3,000 nymphalid species worldwide and some 150–160 in North America. The caterpillars are mostly spiny and feed on a wide variety of host plants. Chrysalides are usually thorny or angular in appearance and hang upside down from silken pads.

Buckeye: *Junonia coenia* — Fig. 15

The characteristic features of buckeyes are their large eyespots with iridescent blue and lilac irises. The species is wide-ranging throughout most of North America in the summer, but it is not able to winter very far north. In Florida it is sometimes confused with the Caribbean buckeye (see below), a similar species that some authorities regard as synonymous with *J. coenia*. Buckeyes usually fly low over the ground, at a height of around 12 to 18 inches (Edwards and Richman 1977).

Caribbean buckeye: *Junonia evarete* — Fig. 16

Southward of a line from Fort Myers on the west coast and Orange County on the east coast there exists another species of buckeye butterfly, one with much smaller eyespots on its wings. This is the Caribbean or West Indian buckeye, a species more common in the Caribbean, whose larvae feed on black mangrove. Although the two species of buckeye may co-occur in south Florida, the Caribbean variety seems largely to replace *J. evarete* in the Keys during summer.

White peacock: *Anartia jatrophae* — Fig. 17

Another south Florida resident, the white peacock is present from the Tampa area southward, except in cold weather. It is more common in swampy or wet regions, where its larvae feed on water hyssop (*Bacopa monnieri*) and ruellia (*Ruellia occidentalis*), but it is also seen in flight over disturbed areas. The Everglades are a good hunting ground for white peacocks.

Malachite: *Siproeta stelenes* — Fig. 18

A glorious butterfly, black with green bands and spots above, marbled green below, the malachite is typically a tropical species from the West Indies and Central and South America, but it is also reported from Dade County and the Florida Keys. More sightings have been noted in recent times, and hundreds were seen in the Homestead vicinity in 1979 feeding on rotting fruit (Lenczewski 1980). In the West Indies larvae feed on *Blechum brownei*, a member of the Acanthaceae (Stiling 1986).

Red-spotted purple: *Basilarchia astyanax* — Fig. 19

At first glance, the red-spotted purple appears much like a swallowtail (especially a pipevine or dark tiger swallowtail), because it is large and has a dark color with blue iridescence on the hindwings. It also has brick-red spots on the underwing. The species is in fact a nymphalid and is thought to mimic the pipevine swallowtail so as to gain protection from the latter's bright coloration, which warns of toxicity. A prerequisite of mimicry is of course that model and mimic be common over the same range and in the same habitat, so, like the pipevine, the red-spotted purple is common throughout most of the eastern United States, ranging from open woodlands to forest edges. It is more common in the northern part of the state.

Viceroy: *Basilarchia archippus*   Fig. 20

A master of mimicry, the adult viceroy gains protection from its similarity to the distasteful monarch, which most birds ignore. The chrysalis mimics a bird dropping, and the young hibernating caterpillar hides among leaves on willow hosts. It is abundant throughout the state but is smaller than the monarch and has heavier black lines on the wings and an additional black line crossing the hindwing. In south Florida, viceroys more commonly mimic the queen butterfly instead (Lenczewski 1980).

Florida purplewing: *Eunica tatila*   Fig. 21

The Florida purplewing is instantly recognizable by the blue-purple iridescence on the upper surface of the wings. No other Florida butterfly has such a metallic sheen (Stiling 1988*b*). For this color to be visible, the butterfly must be in sunlight; otherwise it looks merely brown. The undersides of the wings are bark-colored, and purplewings are well camouflaged at rest on tree trunks. Purplewings are found only in the hardwood hammocks of Dade and Monroe counties and are more common in the Florida Keys (Lenczewski 1980).

Ruddy daggerwing: *Marpesia petreus*   Fig. 22

At first glance the long tails of the ruddy daggerwing make it look like a swallowtail, although the coloration is similar to that of the Julia. The daggerwing can be found from Fort Lauderdale southward, most commonly in May–July but also in October. Hardwood hammocks and thickets in the Everglades are a good place to search.

Variegated fritillary: *Euptoieta claudia*   Fig. 23

This species is found throughout the state, though less commonly in the lower Florida Keys. The caterpillars, white with red bands and black spines, eat an extraordinarily wide range of host plants. The adults have coloration typical of a large number of fritillaries—that is, tawny brown above but with zigzag black lines and whitish brown below. The species frequents open areas such as fields and grasslands.

Pearl crescent: *Phycoides tharos*   Fig. 24

Common throughout Florida for much of the year, this small species is found in open spaces, fields, roads, and streamsides. Present in most of the United States, the pearl crescent is noticeable because of its commonness and the male's habit of darting out to investigate any passing form, including human.

Question mark: *Polygonia interrogationis*   Fig. 25

This butterfly is named after the silvery design on the underside of the hindwings: a distinct silvery comma with an offset dot, forming a question mark. Without the dot, the butterfly would simply be a comma (*P. comma*), but this species only gets as far south as central Georgia, whereas the question mark can be found in most of Florida. No other species of the genus extends this far south. Adults love sap and rotting fruit and can actually become intoxicated if the fruit they are drinking has fermented in the sun.

Painted lady: *Vanessa cardui*   Fig. 26

Truly a cosmopolitan species, the painted lady is found in Africa, Europe, Asia, and North America; there are few areas where there is not some likelihood of seeing it. This wide distribution may be due to the very catholic tastes of the caterpillars, which prefer thistle (*Cirsium*) but will feed on a huge number of other composites and mallows. *Vanessa cardui* colonizes more northerly parts of its range from warmer overwintering sites in the south.

15. Buckeye
Junonia coenia

16. Caribbean buckeye
Junonia evarete

17. White peacock
Anartia jatrophae

18. Malachite
Siproeta stelenes

19. Red-spotted purple
Basilarchia astyanax

20. Viceroy
Basilarchia archippus

21. Florida purplewing
Eunica tatila

22. Ruddy daggerwing
Marpesia petreus

23. Variegated fritillary
Euptoieta claudia

24. Pearl crescent
Phycoides tharos

25. Question mark
Polygonia interrogationis

26. Painted lady
Vanessa cardui

Whites and Sulphurs: Pieridae

Pierids, comprising almost 2,000 species worldwide, are some of the most abundant butterflies in all regions. The brimstones, yellows, or sulphurs are notable migrants and may assemble on damp soil or mud to drink, providing quite a spectacle. The larvae of several "whites" are of economic importance as pests, especially on crucifers (whites) and legumes (sulphurs). Caterpillars are usually smooth, green, and cylindrical. North American species, of which there are about 50–60, are medium-sized butterflies with wingspans between 3 and 6 centimeters. The word butterfly is probably derived from a member of this family: a butter-colored fly. The colors are based on pigments rather than structural separation of light. In the case of the whites, it is derived from a waste product of the insect's metabolism incorporating uric acids.

Florida white: *Glutophrissa drusilla*   Fig. 27

This species is found consistently throughout the year in Dade and Monroe counties, though only rarely. It is more commonly a West Indian and South American species ranging to southern Brazil. It was recently separated from similar oriental *Appias* spp. into the genus *Glutophrissa* (two species). Males are brilliant white on both surfaces; females are more variable, with off-white coloration, sometimes with dark borders to the forewing.

Great southern white: *Ascia monuste*   Fig. 28

Numerous toward coastal areas of the state, the great southern white is slightly larger than the Florida white and has dark tips to the wings. Some females may be entirely suffused with smoky scales, more commonly in high summer. These whites undergo large-scale migrations, probably when local food supply is scarce, and during such times they may be seen in abundance along waterways of the east coast (Stirling 1923, Fernald 1937).

Sleepy orange: *Eurema niccipe*   Fig. 29

A small sulphur, bright gold-orange above and with an uneven black border, the sleepy orange can be prolific in the South, with adults visible in all months. Indeed the common name may derive from its habit of hibernating through the cooler days of Southern winters. The species has a rapid zig-zag flight. Males are often abundant at mud puddles.

Cloudless sulphur: *Phoebis sennae*   Fig. 30

This large sulphur is almost pure yellow in coloration. The most common large sulphur in the state, it is also distributed through the Gulf states and southern California. Some caterpillars hide within tents formed of silk and leaves of their host plants, normally legumes, but others occur right out in the open on *Cassia obtusifolia*, a favorite food plant. The common name stems from their tendency to fly only on sunny, "cloudless" days. Individuals have been clocked flying at about 8 mph into a 5 mph headwind; they normally fly about three feet above ground level, exhibiting an oscillating flight path (Correale and Crocker 1976). Adults feed on any one of two or three host plants, from June to November, commonly utilizing *Agalinis purpurea* and *Ipomoea* sp. (May 1988).

Orange-barred sulphur: *Phoebis philea*   Fig. 31

The largest of the commonly encountered Florida sulphurs, this species has a wingspan of up to 8.5 centimeters. Its coloration is bright yellow with a broad orange band bordering the hindwing and an orange bar across the center of the forewing. Common only in the southern half of the state, this is typically a tropical species, common in Central and South America. Both *P. philea* and *P. sennae* can often be seen drinking on river banks.

Skippers: Hesperidae

Skippers derive their popular name from a darting flight, quite unlike that of other butterflies. In many ways they can be regarded as primitive lepidopterans, with many similarities to moths, such as stout bodies and the ability to rest with forewings folded flat on each side of the body. The family is a large one with nearly 250 North American species. Most are small, with subdued colors of browns and occasional lighter markings. Caterpillars are commonly green or whitish and often feed on grasses, weaving silk and leaf shelters during the day.

Silver-spotted skipper: *Epargyreus clarus*   Fig. 32

The silver-spotted skipper is dull brown but with hindwings exhibiting a large irregular patch of silver underneath. It is common throughout the state during most of the year and is a very wide-ranging species both in terms of absolute area occupied, most of North America, and in terms of habitat—parks and gardens to forest.

Long-tailed skipper: *Urbanus proteus*   Fig. 33

This species is another robust dull-brown butterfly but has an intense bluish-green sheen on the base of the wings and on the head and thorax. Specimens characteristically possess tails on the hindwing, about 1.5 centimeters long. This skipper ranges from South Carolina through the Gulf states. It is common from gardens to meadows and is especially evident when the sun shines. The larvae are sometimes a pest on beans.

Tropical checkered skipper: *Pyrgus oileus*   Fig. 34

This species is prevalent in the southern portion of the state from Tampa and Gainesville south, especially in hot arid areas. The black and white checkered coloration is characteristic. Encountered throughout the Caribbean and much of South America, it sometimes becomes extremely abundant.

27. Florida white
    Glutophrissa drusilla

28. Great southern white
    Ascia monuste

29. Sleepy orange
    Eurema niccipe

30. Cloudless sulphur
    Phoebis sennae

31. Orange-barred sulphur
    Phoebis philea

32. Silver-spotted skipper
Epargyreus clarus

33. Long-tailed skipper
Urbanus proteus

34. Tropical checkered skipper
Pyrgus oileus

Hairstreaks, Blues, and Coppers: Lycaenidae

Well represented in all world regions, the family Lycaenidae contains several thousand, generally small, butterflies. About 100 occur in North America, falling into four general groups. The hairstreaks are so called because of the delicate lines on the underside of the wings, quite visible when the butterfly is resting with wings together. These are the commonest members of the family in the South. The blues and coppers, named for their general respective colors, are more common in the North. Lycaenid caterpillars often feed on flowers, buds, or the fruit of plants. Some are protected by ants in return for sticky secretions of honeydew exuded from special glands.

Atala: *Eumaeus atala* Fig. 35

The atala is a beautiful example of Florida's Lepidoptera (Castner 1986) and one that is unique to the area, because no other state can boast of this species. Once common in Dade and Monroe counties, it was thought to have gone extinct in the 1950s, having failed to survive the ravages of collectors and real estate developers. It is now known from Broward and Dade counties in a few isolated areas where its only host plant, coontie (*Zamia pumila*), grows (Baggett 1982, Landolt 1984).

Red-banded hairstreak: *Calycopis cercops* Fig. 36

Perhaps the most common hairstreak in the state, the red-banded hairstreak is common throughout the Southeast. In common with many hairstreaks, it has fairly long tails, which it often rubs together, giving the illusion of a false head. This behavior may have considerable survival value as protection against predators, which strike toward the tail, allowing the butterfly to escape.

Miami blue: *Hemiargus thomasi* Fig. 37

There are at least half a dozen "blues" that can be found in the state of Florida, and this one, the Miami blue, is fairly typical of them. It is found from the Tampa area southward and can even become common in the Keys. Its appearance changes somewhat with the seasons; black encroachment from the wing margins is more common in summer and fall, whereas in winter and spring individuals appear more blue.

35. Atala
    Eumaeus atala

36. Red-banded hairstreak
    Calycopis cercops

37. Miami blue
    Hemiargus thomasi

Ringlets: Satyridae

Even to many lepidopterists, the dull brown color of most satyrids is uninteresting; it gives rise to their alternative name of browns. The name ringlets is derived from the many small eyespots on the underside of the wings. The family is a large one, with about 3,000 species worldwide and 50 in North America. Many tropical satyrids fly at dusk, but our temperate species do not share this habit. Most do not fly far, simply flitting within a small area.

Large wood nymph: *Cercyonis pegala*   Fig. 38

A large ringlet, with wingspan between 5 and 8 centimeters, the large wood nymph has wings of a light to dark brown with one or two large eyespots above and below and up to six smaller ones below. It is more common in the northern half of the state and is encountered in open oak or pine woodlands during mid and late summer. This is the only wood nymph (*Cercyopnis*) east of the Mississippi.

Milkweed Butterflies: Danaidae

The Danaidae, predominantly a tropical and subtropical family of butterflies, are also referred to as tigers, in reference to their striking orange coloration with severe black stripes. This pattern is an advertisement of noxious taste, the noxious element consisting of heart poisons (cardiac glycosides) synthesized from their food plants (Asclepiadaceae and Apocynaceae) by the larvae and stored by the adults. The reinforcement of the same basic warning coloration by many species of danaids is known as Mullerian mimicry and tends to deter predators from attacking any prey species displaying it (cf. the black and yellow banding patterns of many different wasp species). Although there are about 300 species of milkweed butterflies worldwide, only two are commonly found in Florida.

Monarch: *Danaus plexippus*   Fig. 39

The monarch is a large butterfly, with an 8- to 10-centimeter wingspan, bright orange wings with black veins, and no black line across the hindwing. The similar viceroy is smaller and has a black line across the hindwing (Fig. 20). Often seen on the wing in October, the monarch undertakes long migratory journeys from northern states to overwintering sites in Mexico, passing through Florida on the way (Urquhart 1976, Nagano and Freese 1987). Monarchs cannot tolerate cold weather, although Florida may be sufficiently warm for some communal overwintering roosts to form. Caterpillars are off-white with black and yellow transverse stripes and have a pair of black filaments extending from the front and rear ends. Normal larval hosts are poisonous milkweeds, which give larvae and adults a poisonous taste. Recently some alternative hosts have been found, such as dogbane (*Apocynum*), which contain no noxious chemicals. Adults hatched from larvae feeding on these plants are not poisonous, they therefore rely solely on their resemblance to noxious individuals (automimicry) for protection. The monarch's strong flight ability probably enhances its almost circumtropical distribution; it can be found in North and South America, the Canary Islands, Indonesia, and Australia. Some monarchs tagged in Ontario have been known to fly 1,870 miles in 129 days on their journey southward to Mexico. Populations of monarchs in the western United States may overwinter on the coast between Monterey and San Diego. For example, John Steinbeck refers to overwintering roosts at Pacific Grove in his novel *Sweet Thursday*. In the spring monarchs make the return journey north so that larvae can utilize the abundant milkweeds that grow in the United States.

Queen: *Danaus gilippus*   Fig. 40

Possibly more abundant in Florida than the monarch is the closely related queen butterfly. It is also large but has a deep brown base coloration (as opposed to orange) with black margins and *fine* black veins. The queen is nonmigratory and, being unable to withstand cold winters, is normally a resident only of the Gulf Coast states.

38. Large wood nymph
   Cercyonis pegala

39. Monarch
   Danaus plexippus

40. Queen
   Danaus gilippus

41. Tulip-tree silkmoth
   Callosamia angulifera

42. Luna moth
Actias luna

43. Io moth
Automeris io

44. Polyphemus moth
Antheraea polyphemus

45. Spiny oakworm
Anisota stigma

46. Rosy maple moth
Dryocampa rubicunda

Giant Silkworm Moths: Saturniidae

Silkworm and emperor moths are medium to large in size and include the largest moths in our area (up to 15 cm wingspan). They have a stout, densely hairy body and a small head with pectinate (branched) antennae. In some the proboscis is absent: they do not feed. Larvae are usually fleshy and armed with spines or hairs, which may contain poison. These feed on a wide variety of trees and shrubs and usually pupate in a well-built silken cocoon. The commercial silkworm moth (*Bombyx mori*) is, however, not native to North America and belongs to another family (Bombycidae). The origin of the commercial silkmoth is thought to be in China, though there are no native populations remaining. Silk production was kept a closely guarded secret in China, and revealing this secret was punishable by death. Eventually, in the 6th century, two monks smuggled eggs out to the West in their holy bamboo sticks.

Tulip-tree silkmoth: *Callosamia angulifera*   Fig. 41

A large dark silkmoth confined to the northern border regions of Florida. Similar species, also confined to the northern counties, especially the northwest, include the very similar Promethea moth (*C. promethea*) and the even rarer and beautiful cecropia moth (*Hyalophora cecropia*), unmistakable with its red body and red-bordered wings of grey (Covell 1984), but normally a denizen of northern climes.

Luna moth: *Actias luna*   Fig. 42

This spectacularly beautiful moth is pale green with eyespots (moons) on each wing and long sweeping tails on the hindwings. Almost as common around lights as the polyphemus moth, it is seen from March onward but is more prevalent in the northern counties of Florida. Similar moon moths are found in Asia and Africa. There are two distinct peaks of abundance, one in spring and one in the fall.

Io moth: *Automeris io*   Fig. 43

The io moth is the smallest of our large saturnids and the only one with large, characteristic "bull's-eyes" on the hindwings. It is common statewide, generally from March to November. The larvae feed on a wide variety of shrubs, including azalea, saw palmetto, avocado, hibiscus, and dogwood. The male has yellow forewings, whereas the less-common female, shown here, has red to brownish coloration. A reddish form occurs in the Keys.

Polyphemus moth: *Antheraea polyphemus*   Fig. 44

The polyphemus is probably the most common large silkmoth in Florida, being found throughout the state from February to July and again from October to December, though it is less common in southern counties (Brown 1972a). Adults come readily to lights. This moth is named after the one-eyed giant Polyphemus of Greek mythology. Larvae are large, green, and fleshy (Figure 67).

Spiny oakworm: *Anisota stigma*   Fig. 45

The spiny oakworm is the only member of the genus in which the males can be commonly found at lights. Its flight period is between June and August. The larvae are featured in Figure 77.

Rosy maple moth: *Dryocampa rubicunda*   Fig. 46

A beautiful and delicate-looking moth with a yellow body and largely pink wings, this species is found throughout the state. Larvae, commonly called green-striped maple worms, feed on maple in general and on oaks in Florida and can cause serious defoliation.

Regal moth: *Citheronia regalis*   Fig. 47

There is generally but one brood of this moth, adult males of which can commonly be seen in July and August around convenience-store lights. Females are rarely encountered. The species is more common in the north of Florida but can be found as far south as Lake Okeechobee (Brown 1972b). The larvae (hickory horned devils) are shown in Figure 69. The rarer pine-devil moth, *Citheronia sepulchralis*, has larvae with short horns.

Imperial moth: *Eacles imperialis*   Fig. 48

The imperial moth is another majestic, imposing moth that comes to street lights and other lights all over Florida (except in the Florida Keys). There are generally two broods of adults, which fly from April to July and from August through October. Larvae are large and generally green and hairy, with white spots surrounding the spiracles along the flanks. They feed on a wide variety of trees. In some instances a brown color phase is evident in the caterpillars.

Sphinx Moths: Sphingidae

Sphinx moths, or hawk moths, are medium to very large moths with a very robust, chunky body shape that tapers to a sharp point at the end of the abdomen. In some species, the proboscis is extremely long, and in the pupal stage it is often housed in a special sheath that curls away from the body and is reminiscent of a jug handle. The larvae are hairless, and nearly all have single, prominent horns at the end of the body. Most sphinx caterpillars pupate inside leathery cases in the soil. Some adults can be collected over flowers at dusk, others fly later at night, and a few can be found on flowers in the day. They are strong fliers, and their wings beat rapidly; for this reason some day fliers resemble hummingbirds or large bees.

Pink-spotted hawk moth: *Agrius cingulatus* Fig. 49

The pink-spotted hawk moth or sweet potato worm occurs statewide and is relatively common. It is the only sphinx in our area with pink crossbars on the abdomen. The hindwings are grey with black bands and some pink shading toward the base. This wide-ranging species is found from tropical South America through to the southern United States.

Rustic sphinx: *Manduca rustica* Fig. 50

The rustic sphinx is one of the largest sphinx moths in our area, with a wingspan of up to 15 cm. (The largest is the giant sphinx, *Cocytius antaeus*, wingspan 13-18 cm, found in the Keys and south Florida.) Rustics can be found all over the state. Their larval foods include vines (*Bignonia* sp.) and fringe tree (*Chionanthus virginicus*).

Catalpa sphinx: *Ceratomia catalpae* Fig. 51

The catalpa sphinx is found mostly in north Florida. The larvae may be discovered in groups feeding on their only host plant, the catalpa tree, which they can, at times, defoliate. Larvae occur in two color forms; the more common is black on top and pale yellow underneath.

Twin-spotted sphinx: *Smerinthus jamaicensis* Fig. 52

This species is found only in the north of the state. The distinguishing feature of this sphinx is that the blue spots on the hindwing are divided by a black bar to form two spots. A third small blue spot is present toward the base of the hindwing in some specimens.

Blinded sphinx: *Paonias excaecatus* Fig. 53

Larvae of the blinded sphinx are light green, studded with pointed granulations. There are seven oblique, yellowish stripes running backward on each side of the body. The spiracles are lilac or black. Larvae feed on *Prunus* species. Adults have a strongly scalloped outer wing margin.

Small-eyed sphinx: *Paonias myops* Fig. 54

This is primarily a northern species, smaller than the blinded sphinx (above). The outer margins of the forewing are doubly indented. The larvae have rose-colored spiracles; otherwise, they are like those of the blinded sphinx.

47. Regal moth
Citheronia regalis

48. Imperial moth
Eacles imperialis

49. Pink-spotted hawk moth
Agrius cingulatus

50. Rustic sphinx
Manduca rustica

51. Catalpa sphinx
Ceratomia catalpae

52. Twin-spotted sphinx
Smerinthus jamaicensis

53. Blinded sphinx
Paonias excaecatus

54. Small-eyed sphinx
Paonias myops

55. Banded sphinx
Eumorpha fasciata

56. Hydrangea sphinx
Darapsa versicolor

57. Hog sphinx
Darapsa myron

58. Tersa sphinx
Xylophanes tersa

Banded sphinx: *Eumorpha fasciata* — Fig. 55

A tropical species, *E. fasciata* is common statewide, including the Florida Keys. Larvae feed on members of the Onagraceae family, especially on primrose willow, *Ludwigia peruviana*.

Hydrangea sphinx: *Darapsa versicolor* — Fig. 56

Another sphinx that has been recorded from Pensacola to Key West, this moth may vary from uncommon to locally common. As the name suggests, larvae prefer to feed on wild hydrangea, *Hydrangea arborescens*. They will also take button-bush (*Cephalanthus*).

Hog sphinx: *Darapsa myron* — Fig. 57

The hog sphinx is also known throughout some of its range as the Virginia creeper sphinx. In Florida, caterpillars like to feed on grapes and on pepper vine (*Ampelopsis arborea*). The species is common all over the state, probably in every month.

Tersa sphinx: *Xylophanes tersa* — Fig. 58

Xylophanes tersa is easily recognized by its long pointed abdomen and by the jagged black markings on the hindwings. The forewings are pale and uninteresting. It is a common species statewide, with flight activity recorded from February to November, but only in the summer months in north Florida.

Tiger Moths: Arctidae

This family contains not only orange or yellow "tiger moths" with brown or black markings but also some species with cryptically colored forewings and flashy underwings. There are about 10,000 known species in the world. Many fly by day, but most are nocturnal. The larvae are usually extremely hairy and are known as "wooly bears" or tussocks (see Figures 73 and 76). Most are partial to poisonous varieties of plant, whose toxins are utilized by the caterpillars as part of their own defense. Such behavior also imparts a toxicity to the adults, who advertise this fact with bright coloration. Recent experiments have shown some species advertise their distastefulness to night-flying bat predators by generating their own high-frequency sounds, which the bats can detect and which they learn to associate with bad taste.

Rattlebox moth: *Utethesia bella* — Fig. 59

The rattlebox or bella moth is a conspicuous day flier, with orange and black forewings. It often seems to disappear in flight, alighting on grass and wrapping its wings around the blade, thus hiding the distinctive pink coloration of the hindwings, so evident in flight. It is particularly common around legumes, especially rattlebox, which the larvae feed on, frequently penetrating the "rattles" themselves. A substance poisonous to vertebrates has been isolated from the secretions of the rattlebox moth thoracic glands. Mixed with the insect's haemolymph (blood and lymph), this is ejected in a froth, with an unpleasant smell, as a defensive reaction.

Giant leopard moth: *Ecpantheria scribonia* — Fig. 60

Common statewide, this moth is easily recognized by the white wings with bluish-black circles on the forewing. The typical form of the male in Florida is *denudata*, whose scales on the apical fifth of the wings appear to have been rubbed off. The large black and hairy larvae are often garden pests, feeding on cabbages.

Owlet Moths: Noctuidae

With over 20,000 species worldwide and 2,900 in North America, this is the largest lepidopteran family. The vast majority of the species are small and dull, with lightly patterned greys and browns predominant. Many of the small moths attracted to outside lights are noctuids. Larvae feed in a wide variety of situations, chewing on foliage and boring stems, roots, and fruit. Many are serious economic pests of cultivated crops and trees.

Spanish moth: *Xanthopastis timais* Fig. 61
This atypically colorful noctuid has a hairy black body and black markings on a pink forewing. It is common in Florida throughout most of the year, and its larval food includes spider lily.

Underwing: *Catocala* sp. Fig. 62
There are many species of underwing in the state; Kimball (1965) records 36. With the wings closed, the underwings merge into the background like those of typical noctuids. The hindwings, however, are usually a splash of red, orange, or white with black peripheral bands. It is thought that, when camouflage fails and predators locate these moths, as a last resort *Catocala* flash their bright underwings in an effort to startle their enemies, from which escape may then be possible.

Caterpillars

Caterpillars are the larvae of the Lepidoptera, the moths and butterflies. It is often the case that a particularly beautiful butterfly has an undistinguished and easily overlooked caterpillar, whereas the more visible and interesting caterpillars usually assume the guise of dull brown moths. Because it is the intention of this volume to illustrate the more interesting and visually obvious insects, the more commonly encountered caterpillars are allotted a section of their own. It is worth noting that, of all the caterpillars likely to be found in Florida, only 27 are provided with stinging hairs, mainly in the family Limocodidae. Three species often encountered are the saddle-back (*Sibine stimulea*), the larva of the hag moth (*Phobetron pithecium*) and the larva of the puss moth (*Megalopyge opercularis*) (Anonymous 1921, 1945). The results of the irritating hairs from these caterpillars vary greatly with the individual stung. Puss moth larvae, however, can cause severe inflammation and swelling.

59. Rattlebox moth
Utethesia bella

60. Giant leopard moth
Ecpantheria scribonia

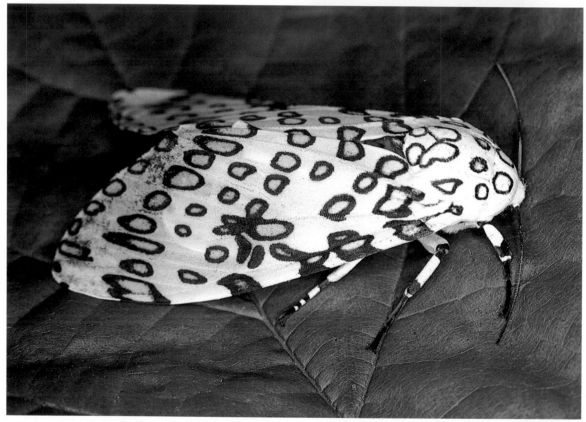

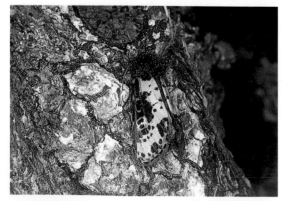

61. Spanish moth
Xanthopastis timais

62. Underwing
Catocala sp.

63. Eastern black swallowtail caterpillar
Papilio polyxenes

64. Orange dog caterpillar
Heraclides cresphontes

65. Brazilian skipper caterpillar
Calpodes ethlius

66. Gulf fritillary caterpillar
Agraulis vanillae

67. Polyphemus caterpillar
Antheraea polyphemus

Eastern black swallowtail caterpillar: *Papilio polyxenes*   **Fig. 63**

About 5 centimeters long when mature, this caterpillar is perhaps the most commonly encountered of all the swallowtail larvae (except perhaps those of the orange dog; see below). Eastern black swallowtail caterpillars feed on members of the carrot family (Apiaceae) and, as a result, can be found in the back yard on the parsley or celery making pests of themselves. They will also feed on some citrus. Wild hosts include water dropwort, *Oxypolis filiformis*.

Orange dog caterpillar: *Heraclides cresphontes*   **Fig. 64**

Known as the "orange dog" by citrus growers, larvae of the giant swallowtail are rarely considered a pest on oranges and only infrequently subject to spraying on young plants. Caterpillars can also be found on hop tree, *Ptelea trifoliata*. Larvae of H. *cresphontes* (and of some other swallowtails) resemble large bird droppings, with dirty buff patches and saddles.

All swallowtail larvae have a pair of forked scent horns, called an osmeterium, that are normally concealed just behind the head, but can be everted as shown here. In the giant swallowtail they are usually red. They give off a powerful odor not too dissimilar from that of citrus, on which the animal is reared. These scent horns will be flashed when the insect is disturbed and are thought to deter enemies.

Brazilian skipper caterpillar: *Calpodes ethlius*   **Fig. 65**

Two species of caterpillar are likely to be the culprits when damage to cannas (*Canna flaccida* and C. *indica*) is found in the yard. One is a small caterpillar, *Geshna cannalis*, that rolls the leaves and skeletonizes them and that is normally found gregariously. The other, larger individuals are Brazilian skippers or larger canna leaf rollers, which take big chunks out of the leaves. The chrysalis of the Brazilian skipper is a ghostly green and is tucked away in a leaf roll.

Gulf fritillary caterpillar: *Agraulis vanillae*   **Fig. 66**

This species is the most common caterpillar to be found on passion flowers. When ready to pupate, caterpillars leave the host plant and relocate, usually to a solid object like a fence, or gate, or plant stem, often some distance from the host. Some swallowtail larvae also travel great distances to pupate. The chrysalis of the Gulf fritillary is about 2.5 cm long, mottled brown and warty, and resembles a dried-up leaf.

Polyphemus caterpillar: *Antheraea polyphemus*   **Fig. 67**

This is the most commonly encountered large caterpillar in the north of the state and can be found on a wide range of trees and shrubs. The bright, fleshy green larvae retract their heads and remain motionless when disturbed. The larvae feed on a wide variety of hosts such as oaks, ashes, maples, birches, and hickories. In winter, after leaves have fallen, cocoons are highly visible, as they hang by silken strands from twigs.

Luna moth caterpillar: *Actias luna*   **Fig. 68**

Larvae feed on a variety of host plants; sweet gum is commonly utilized in Florida. The pupa remains active in its papery cocoon, which is usually spun on the ground. Because the cocoon is concealed among leaf litter, it is hard to find in winter.

Hickory horned devil: *Citheronia regalis*   **Fig. 69**

The hickory horned devil is an imposing beast, with beautiful aquamarine coloration and frightening red horns. Although fearsome in appearance, they cannot sting. When touched, these caterpillars will swing their horns around, but fear not; they are not poisonous or irritating. The caterpillar feeds on a wide variety of trees and may also be encountered, full-grown, on the ground as it searches for a pupation site. Devils can easily be reared on pecan trees.

Tobacco hornworm: *Manduca sexta*   **Fig. 70**

This caterpillar is commonly encountered by the gardener as a pest of tomatoes and potatoes. Its characteristic features are a red-tipped horn at the tip of the abdomen and seven oblique white lines along each side of the body. Another, very similar caterpillar that feeds on the same plants is the tomato hornworm, the larva of *Manduca quinquemaculata*, the five-spotted hawk moth. This caterpillar has eight L-shaped lines on the sides and a black-edged green horn. Adults of the tobacco hornworm, *Manduca sexta*, are common between May and October.

Eastern tent caterpillar: *Malacosoma americanum* Fig. 71

Eastern tent caterpillars, members of the family Lasiocampidae, are easily recognized by the extensive white webbing (tent) that the gregarious larvae spin in the forks of various trees and shrubs. These are serious defoliating pests. The larvae hide in the webbing during the day and come out to feed at night. When full grown, caterpillars leave the host tree and disperse to pupate singly. Common only in the north of the state from Orlando northward.

Bagworm: *Thyridopteryx ephemeraeformis* Fig. 72

These larvae form characteristic spindle-shaped silken cases covered with bits of leaves and twigs and retain the case throughout their lifetimes, enlarging it with each moult. The caterpillars pupate inside, and females, which are wingless, never leave, attracting males by pheromones (chemical sex attractants). The male thrusts his abdomen through the open lower end of the case to mate. Eggs are laid inside the case, and hatched larvae eventually crawl away to feed and construct their own homes. Several other species of this family, Psychidae, occur in Florida.

White-marked tussock moth caterpillar: *Orgyia leucostigma* Fig. 73

With a characteristic "toothbrush-like" appearance, this caterpillar is very catholic in its tastes, with over 140 known hosts. It is especially common on live oak in the northern half of the state from March onward. Tussock moths belong to the family Lymantriidae, which also contains the infamous gypsy moth, *Lymantria dispar*, an extremely serious pest in hardwood forests from Nova Scotia to North Carolina. The gypsy moth was originally introduced into this country from Europe in 1868 by Leopold Trouvelot, who hoped to raise larvae for silk production. Unfortunately, some of his moths escaped, and the dreadful damage of their unchecked populations began to spread from the Boston area outward.

Pale tussock moth caterpillar: *Halysidota tessellaris* Fig. 74

These larvae can be found on a wide array of deciduous trees and shrubs; this one was feeding on water oak, *Quercus nigra*. The larvae have distinctive hair pencils rising in pairs from the second and third thoracic and eighth abdominal segments. They are usually found singly from August to October.

Yellow-necked caterpillar: *Datana ministra* Fig. 75

The yellow-necked caterpillar, a member of the Notodontidae, is commonly seen feeding in groups on leaves. Two color phases are recognized, a red and yellow form and a black form. Both lift front and rear parts of the body, then stay motionless when disturbed. They are common on a wide variety of trees statewide.

Red-humped caterpillar: *Schizura concinna* Fig. 76

Another polyphagous notodontid, the red-humped caterpillar may, like the yellow-necked caterpillar, cause damage to host plants that include azalea, rose, camellia, red bay, and redbud. Like most of the caterpillars described here, the larvae are more commonly seen than the adults. When at rest, the larva holds its rear end in an elevated position, and, when handled, it gives off a pungent, disagreeable odor.

Spiny oakworm: *Anisota stigma* Fig. 77

This small saturniid species is more common in the north of the state, concurrent with distribution of the oaks on which it normally feeds. It is often found feeding gregariously on water oak branches, which it may completely strip of leaves.

Saddleback caterpillar: *Sibine stimulea* Fig. 78

This limacodid larva is brownish except for a green patch, which resembles a saddlecloth, on the middle of its back. It body is armed with fascicles of poisonous spines, and it also has a pair of spiny tubercles at each end. Found on a wide variety of trees and ornamental plants, including dogwood, oaks, and citrus.

Hag moth caterpillar: *Phobetron pithecium* Fig. 79

Hag moths are so named because the third, fifth, and seventh pairs of lateral processes (of which there are nine) are long, curved, and twisted, suggestive of the disheveled locks of a "hag." These processes are clothed with stinging hairs. The caterpillar also goes by the alternative name of monkey slug. It feeds on trees such as dogwoods, hickories, and oaks and on shrubs and citrus.

68. Luna moth caterpillar
    Actias luna

69. Hickory horned devil
    Citheronia regalis

70. Tobacco hornworm
    Manduca sexta

71. Eastern tent caterpillar
    Malacosoma americanum

73. White-marked tussock moth caterpillar
    Orgyia leucostigma

72. Bagworm
    Thyridopteryx ephemeraeformis

74. Pale tussock moth caterpillar
Halysidota tessellaris

75. Yellow-necked caterpillar
Datana ministra

76. Red-humped caterpillar
Schizura concinna

77. Spiny oakworm
Anisota stigma

78. Saddleback caterpillar
Sibine stimulea

79. Hag moth caterpillar
Phobetron pithecium

$\mathcal{D}$ragonflies and damselflies: order ODONATA

Members of the Odonata represent the most ancient of flying insects, the dragonflies, having first appeared on earth in the Carboniferous period 350–280 million years ago. The word Odonata comes from the Greek word meaning tooth. The reason for this class name of dragonflies and damselflies is unclear, but it may refer to the toothy jaws or mandibles. Nevertheless, because the immature or larval stages of these insects develop in water, Florida, with its abundance of lakes and slow-moving rivers has a profusion of species. At least 120 different dragonflies occur in Florida, 60% of Nearctic or temperate origin from the north, 23% of neotropical origin from South America and the Caribbean, and 17% endemic (Byers 1930). These figures reflect the subtropical nature of much of Florida's climate and habitats. Most dragonflies appear here in April, at least a month earlier than their more northern counterparts. Many dragonflies in Florida are large, with a wingspan of 5–9 cm. They cannot fold their wings back and are therefore at their most active in the air rather than on the ground. Damselflies, on the other hand, can fold their wings back over their bodies at rest. They are also smaller and more delicate than dragonflies. Both types of insect are predaceous both as adults, skillfully catching other insects on the wing, and as larvae, feeding underwater on other freshwater organisms. Adult males commonly patrol a territory along a stretch of shoreline that includes suitable egg-laying sites. A male normally allows only females with which he has recently mated to oviposit on his patch. To cool off in the full heat of the day, perching males adopt the obelisk stance, pointing the abdomen straight up toward the sun to minimize the surface area available to the sun's hot rays.

Brown-spotted yellow-wing or halloween pennant: *Celithemis eponina*   Fig. 80

A common dragonfly in Florida from mid-April onward, this species frequents the borders of ponds with weedy bottoms. It has a distinct appearance, with yellow wings banded and spotted with brown. The head and body are amber. This is the largest member of a genus with 10 species, all of which occur east of the Rocky Mountains.

Green clearwing or eastern pondhawk: *Erythemis simplicicollis*   Fig. 81

Active nearly year-round in the south, this bright green dragonfly is a common sight resting on bare ground and vegetation near the borders of clear lakes. It captures a wide variety of prey on the wing, including damselflies. Particularly inquisitive adults will also settle near or even on human passersby.

Swift long-winged skimmer or blue dasher: *Pachydiplax longipennis*   Fig. 82

The short abdomen of this species makes the wings appear extra long. It is found in the vicinity of broad streams and large ponds or lakes. Males will often engage in aerial dogfights on meeting; females rest on trees away from the shore. When laying eggs, they fly close to the water surface, flicking the abdomen downward to wash off eggs.

Comet darner: *Anax longipes*   Fig. 83

This species is one of the fastest and biggest of the common dragonflies. Another giant is the common green darner, *Anax junius*, whose thorax is green but whose abdomen can be blue. Naiads, the aquatic immatures of *Anax* and other dragonflies, can feed on tadpoles and small fish, in addition to aquatic insects. Found near large ponds and streams.

Doubleday's bluet or Atlantic bluet: *Enallagma doubledayii*   Fig. 84

A variety of delicate blue damselflies inhabit Florida, and this bluet is perhaps the most common around sandy-bottomed ponds, especially in pine barrens. During mating the male holds the female by her neck with pincers at the tip of his abdomen, while she rests on plants above the water surface.

Black-winged damselfly or ebony jewelwing: *Calopteryx maculata*   Fig. 85

These beautiful insects have brown to black wings and flit gently around the edges of slow streams in forests. The male's body is a striking metallic green, the female's a duller green. Females insert their eggs into the stems of aquatic plants.

Variable dancer: *Argia fumipennis*   Fig. 86

Several violet-colored damselflies can be found in Florida, and this is the most common species in the genus. Variable dancers can often be seen flying in tandem over streams and ponds. Only the males have violet-colored bodies; females are brown to black. When eggs are deposited, the male takes off like a helicopter, lifting the female from the water.

80. Brown-spotted yellow-wing
Celithemis eponina

81. Green clearwing
Erythemis simplicicollis

82. Swift long-winged skimmer
Pachydiplax longipennis

83. Comet darner
Anax longipes

84. Doubleday's bluet
Enallagma doubledayii

85. **Black-winged damselfly**
Calopteryx maculata

86. Variable dancer
Argia fumipennis

Grasshoppers and crickets: order ORTHOPTERA

Grasshoppers and crickets are noted for their large hind legs and jumping ability (to escape predators) and also for their singing to attract potential mates (only the males sing). The sound-producing, or stridulatory, apparatus and the ears are the major characteristics of the large order Orthoptera. Crickets' songs are much higher pitched and seemingly more musical than those of grasshoppers. Other distinguishing features of crickets include shorter forewings and three-segmented tarsi or feet. Grasshoppers are conveniently subdivided into the short-horned varieties (family Acrididae), whose antennae or feelers are usually less than half the length of the body, and the long-horned types (family Tettigoniidae), which often possess antennae longer than their own bodies. Orthopterans are generally phytophagous (plant-eating), with large flat-sided heads and big, chewing mouthparts. The order name, Orthoptera, means "straight wing" and refers to the fore pair of wings, called tegmina, which are long and not actively used for flying but sheath the hind pair, which are large and are folded, fanlike, underneath. The females of many species, especially katydids, have swordlike ovipositors, which are used to cut slits into plants, where eggs are deposited. Metamorphosis in these insects is simple; juveniles, on hatching from the egg, already resemble adults but are without wings or genitalia. These develop with successive molts.

Males' songs may have the unfortunate effect of attracting predators as well as females. It has been suggested that certain birds are attracted to taxiing aircraft because the engines sound like some unspecified type of cricket. Cats have certainly been observed to locate singing crickets by their sounds and then to dispatch the singer (Walker 1964), and geckos may be able to perform the same feat.

Southeastern lubber grasshopper: *Romalea microptera*   Figs. 87 & 88

The lubber grasshopper is commonly encountered in Florida in coastal areas, where it feeds vigorously on salt marsh plants. It can be found inland as well. Lubbers have a catholic diet and eat a wide variety of herbs and shrubs. Adults are therefore seen in many situations, for example on highways as they move across the road from one patch of vegetation to another. Unlike most other grasshoppers, the lubber has only short stubby wings and cannot fly, hence its propensity for walking across the road. Although its colors vary somewhat, the lubber is certainly brightly colored, a warning sign for predators. The flightless lubber cannot easily escape attack, but it deters its enemies with a foul-smelling defensive secretion, and its attackers learn to associate this unpleasant experience with the lubber's bright warning colors. The defensive secretions of lubbers show extreme chemical variation even among individuals in the same population feeding on the same foods, and a wide range of predators is probably deterred (Jones et al. 1986). Nymphs are black with a light-colored stripe running down the dorsal surface. They often move from their hatching areas in the sand along definite trails to areas richer in succulent foliage. Densities of up to 75 of the young black hoppers per yard of trail have been counted (Watson 1941).

Bird grasshopper: *Schistocerca alutacea* Fig. 89

Unlike the lubber grasshopper, the bird grasshopper has a powerful flight. It inhabits open sandy woods and grassy areas and merges well with the background vegetation because of its greenish-brown coloration. The female thrusts her eggs into the soil, where they hatch into nymphs in about a week. The genus *Schistocerca* contains some of the most damaging grasshoppers in the world, especially in Africa, where swarms of *Schistocerca*, commonly known as locusts, can devastate crops. These swarms, which may contain 40 billion individuals and weigh 70,000 tons, can eat as much food in a day as the population of New York can eat.

American bird grasshopper: *Schistocerca americana* Fig. 90

Another large grasshopper, brown and beige along the back with black markings, and another strong flyer, the bird grasshopper commonly flies into a tall tree if disturbed. At rest, this grasshopper usually moves to the far side of an object like a twig and draws its hind legs into a jumping position. Active throughout the year except in cool periods, which reduce its movements, it feeds on a wide variety of grasses, herbaceous plants, and trees. Unlike the majority of grasshoppers, which overwinter as eggs, S. *americana* overwinters as an adult and lays eggs in March and April. These hatch in April and May. There are two generations a year; the second or summer generation consists of individuals that hatch during August and September (Kuitert and Connin 1952).

Carolina locust: *Schistocerca damnifica* Fig. 91

This grasshopper is often heard by naturalists because it creates a fast beating sound with its wings during flight. The observant entomologist should also see a flash of yellow-fringed black hind wings as it takes off. When resting the insect is a cryptic cinnamon brown color. It is common along roadsides and in fields throughout the state, feeding on a variety of grasses and herbs. The stridulations of this and the other short-horned grasshoppers are produced by what could be termed the washboard mechanism. The process involves friction between a row of pegs on the inside of the hind leg and one or more pronounced veins on the forewing.

Angular-winged katydid: *Microcentrum retinerve* Fig. 92

Katydids are members of the Tettigoniidae and have extremely long antennae. The common name is supposedly derived from the male's song "katy-did-katy-didn't." The song is produced when specialized veins on the bases of the forewings are rubbed together, essentially a tooth-and-comb technique. The hearing organs are called tympana and are located at the bases of the front tibiae (legs). Most species live in forest trees, where they feed on leaves, but some can cause damage to citrus leaves and fruit (Griffiths 1952a, b). Their green coloration provides perfect camouflage among the foliage.

Field cricket: *Gryllus* sp. Fig. 93

The songs of crickets are distinctive and may be quite irritating when the species is the house cricket, *Acheta domestica*, which lives indoors and calls repeatedly at night. The field cricket is black to dark brown, compared to the house cricket's lighter brown. Both species scavenge for loose plant and animal material on the ground.

Mole cricket: *Scapteriscus* sp. Fig. 94

Mole crickets are one of Florida's major pests, burrowing in the soil and cutting off plant roots. Mole crickets are a dull earth-brown color with no evidence of camouflage or gaudy colors because they are exposed to few predators in their underground existence. The large spadelike forelegs are marvellously adapted to digging through the soil. Mole crickets are sometimes common at certain times of the year on the surface of large grassy areas, such as playing fields. During the mating season they fly and are attracted to lights. The calls of the mole crickets are low-pitched trills or chirps, often amplified by a megaphone-like burrow. There are other species and genera of burrowing crickets in Florida, but the three pest species belong to the genus *Scapteriscus*.

87. Southeastern lubber grasshopper
Romalea microptera

88. Lubber grasshopper nymph
Romalea microptera

90. American bird grasshopper
Schistocerca americana

89. Bird grasshopper
Schistocerca alutacea

91. Carolina locust
Schistocerca damnifica

92. Angular-winged katydid
Microcentrum retinerve

93. Field cricket
Gryllus sp.

94. Mole cricket
Scapteriscus sp.

True bugs: order HEMIPTERA

Although most people commonly refer to all insect "creepy-crawlies" as bugs, scientifically this name is reserved for those insects possessing piercing and sucking mouthparts housed in a long beaklike rostrum. Such mouthparts are used in most species to tap plant sap and in some to suck out the body fluids of other insects or the blood from higher animals. The true bugs have forewings that are leathery at the tips, clearly visible when the wings are folded flat on the back. The other half of the wings is normal, that is membranous, giving rise to the order name Hemiptera or "half wing." There is a closely related group of insects, the Homoptera (see page 00) that also have beaklike rostrums, but in these the forewings are entirely membranous. Some entomologists refer to the true bugs as Heteroptera and call the Homoptera and Heteroptera together the Hemiptera.

The true bugs can swing their beaks forward from the resting position on the underside of the body and thus can exploit food sources other than the plant directly beneath them. Many are predators of other insects. Most true bugs also possess stink glands and advertise their repugnant smell by bright colors. Others use smell only as a secondary line of defense, relying on cryptic camouflage to avoid predators. Metamorphosis in all bugs is through five nymphal instars or developmental stages; each succeeding nymphal instar resembles the adult more closely.

St. Andrew's cotton bug: *Dysdercus andreae* Fig. 95

These bright red bugs are unmistakable, with their white "St. Andrew's Cross" marking on the front wings. They are commonly referred to as "red bugs" or "cotton stainers," because when infesting cotton they infect it with a fungus that leaves a red stain on the cotton fibers, rendering the bolls useless for market. Strictly speaking, however, the most common cotton stainer is actually the closely related *Dysdercus suturellus* (red legs and no white cross). All species of this family (the Pyrrhocoridae) are phytophagous, feeding on low vegetation. The bright red coloration is thought to be a warning to predators. They are often found on hibiscus and sometimes on oranges.

Big-legged bug: *Acanthocephala declivis* Fig. 96

The big-legged bug is the largest plant bug in Florida, being over 2.5 cm in length. It is fairly common on a wide range of food plants and trees. When handled, this bug gives off a foul-smelling secretion, but adults usually fly away if approached. The enlarged femora on the hind legs are biggest in males and may be associated with signalling during courtship (Stiling 1987a). Males are also known to fight using their hind legs, in disputes over defended flowerheads (Mitchell 1980). *Acanthocephala femorata*, a large leaf-footed bug, is slightly smaller, with similar leaflike projections on the legs, and A. *terminalis* is yet smaller, at about 15 mm.

Largid bug: *Largus succinctus* Fig. 97

These bugs, family Largidae, are medium-sized and are usually brightly colored with red or brown and black. They resemble the Pyrrhocoridae in coloration, in habits (both suck plant sap), and in distribution, as both are confined to the Southern states.

Wheel bug: *Arilus cristatus* Fig. 98

This large dark-brown to black bug has a characteristic cogwheel-like crest on its back. It has a bad reputation because it will bite if carelessly handled, and the saliva it injects acts as a poison causing *severe* pain, much worse than a bee sting. It normally feeds on other insects, especially caterpillars. The young are long-legged and awkward-looking.

Southern green stink bug: *Nezara viridula* Fig. 99

Stink bugs get their common name from the copious amounts of defensive secretions they discharge when handled. Even the smallest nymphs can produce a substantial smell. An alternative name is shield bugs, referring to the shieldlike shape of the thorax. Stink bugs are usually phytophagous, tapping plant sap, though a few are predators and will also suck the juices of soft-bodied insect larvae.

Brochymena bug: *Brochymena arborea* Fig. 100

In contrast to the foliage-matching green coloration of green stink bugs, *Brochymena* sp. are a brown-mottled grey, perfect camouflage to match the tree bark they inhabit (Castner 1985). These bugs do not penetrate the thick bark but instead are predatory upon soft-bodied prey.

Giant water bug: *Lethocerus griseus* Fig. 101

The Belostomatidae, a family of aquatic insects, contains the largest true bugs. Their flattened hind legs are used for swimming, and the front pair grasp prey, usually tadpoles, small fish, and other insects. Water bugs suck the fluids from their struggling victims with a powerful beak. Human toes may also be seized and bitten under water; hence the common name "toe-biter." Another given name is "electric light bugs" because, although they live only in freshwater ponds and pools, they are attracted to lights during flight. Males carry developing eggs around on their backs, looking like walking incubators. To obtain air to breathe, giant water bugs raise the tips of their abdomens to the water surface and extend two taillike breathing tubes.

95. St. Andrew's cotton bug
Dysdercus andreae

96. **Big-legged bug**
Acanthocephala declivis

97. Largid bug
Largus succinctus

98. Wheel bug
Arilus cristatus

99. Southern green stink bug
Nezara viridula

101. Giant water bug
Lethocerus griseus

100. Brochymena bug
Brochymena arborea

Cicadas and their relatives: order HOMOPTERA

Cicadas and their kin, leafhoppers, planthoppers, treehoppers, and aphids, to name but a few, all have downward-pointing, immovable beaks, with which they exclusively suck plant sap, and uniformly membranous wings (Homoptera means "similar wings"). Many are plant pests and apart from causing physical damage to the plant they are extremely important vectors of plant viruses and diseases. Aphids (greenfly) and leafhoppers are economically devastating in this respect. Most species are small and numerous, clustering together on plant stems or on the undersides of leaves, where they suck the cells, creating a characteristic white stippling. The plant sap they feed upon is passed out as sticky honeydew, which ants like to drink. Consequently ants attend many species of homopterans, protecting them from predators and stroking them to elicit production of honeydew droplets.

Cicada: *Tibicen* sp.   Fig. 102

At between 2.5 and 6 cm long, cicadas are the giants of the Florida homopterans. They are also the loudest. Many homopterans, even small planthoppers, have now been recognized to produce calling and courtship songs, and cicadas, with their large abdomens, produce a constant zithering during the heat of the day. The sound production method is entirely different from that of grasshoppers. The sound-producing tymbal organs are located in the abdomen, and each tymbal is distorted in and out by muscles rather like a tin lid being clicked in and out. Once again, it is the males that produce the courtship song to attract the females. The young drop from eggs laid in trees and burrow into the soil to feed on roots. In northern states nymphs may take 13 or 17 years to develop fully, emerging en masse at the same time. In warm Florida, however, they appear every year.

Treehopper: *Entylia carinata*   Fig. 103

Entylia carinata is a small and strange-looking member of the membracid family. The pronotum projects far back over the thorax and creates the appearance of a thorn on the plant. Each different species of membracid has different-shaped projections, and each is often specific to a certain shrub on which it feeds. They may be tended by ants.

Glassy-winged sharpshooter: *Homalidisca coagulata*   Fig. 104

Among the family Cicadellidae, the leafhoppers, H. *coagulata* is a large member, being about 1.5 cm long. Some say the name sharpshooter comes from its habit of rapidly running sideways, crablike, around a branch when it is disturbed, whereas others call it the sharpshooter because it leaps rapidly from danger with the speed of a bullet. Some authorities suggest the name comes from the machine-gun-like bursts of "honeydew" it excretes while feeding. These are economically important insects, transmitting two diseases, phony peach and Pierce's disease of grapes. They also attack a wide range of other plants.

Flies: order DIPTERA

True flies have a bad reputation as the inhabitants of filth and the purveyors of disease. The order contains such infamous species as houseflies, deer flies, horseflies, sandflies, and mosquitoes. Florida is often regarded as the mosquito capital of the world, with its abundance of still water where larval mosquitoes develop, and at least 67 species are pests here, to varying degrees. Men, incidentally, are supposedly more attractive to mosquitoes than are women (Gilbert et al. 1966). Sandflies too are particularly annoying in coastal areas. These tiny flies, which develop in the sand, give a bite out of all proportion to their size. Annoying as the fly bites are in themselves, they take on particular seriousness when one realizes that half the clinical cases of disease in the world are transmitted by insects and that flies are by far the predominant carriers. Mosquitoes alone can transmit malaria, yellow fever, filariasis, and dengue fever.

Despite the bad qualities of many members of the order, most flies are harmless, and many have redeeming features such as being important pollinators of flowers and crops. About the only fly popularly thought to have any redeeming features by the general public, however, is the Spanish fly (*Lytta vesicatoria*), whose reputation was enhanced by the Marquis de Sade when he reportedly used candies laced with it to fire the lusts of pretty prostitutes in Marseilles (Durin 1980). Unfortunately for the diptera, these flies are in fact blister beetles, whose chemical defense, cantharadin, is extracted not only as an aphrodisiac but also as a diuretic! Cantharadin, of course, can also cause blisters on human skin, which is how these beetles got their name. Taken internally or absorbed through the skin, cantharadin in high enough doses is toxic to mammals. In Jacksonville and Ocala, thoroughbred horses have died from eating locally grown alfalfa hay that was infested with blister beetles (Zak 1986).

There are at least 90,000 species of fly worldwide, occupying many different habitats. They are the only higher insects in Antarctica. The diversity of fly mouthparts has meant flies can live as blood feeders with piercing stylets, as nectar feeders with tubular mouthparts, or as detritivores with lapping and spongy mouthparts to soak up decaying matter. Their larvae are all legless maggotlike creatures, which can live in a wide range of moisture-providing microhabitats from desert sand to lake bottoms, from snow or hot springs to decaying vegetation. Many are internal parasites of other animals, like snails, other insects, and even humans. The common feature of the adults of all these maggots is that the flight machinery consists of only two wings (hence "Di-ptera"). The hind wings common to most other flying insects are reduced to a small pair of knoblike halteres, which act as gyroscopes for balancing.

Black horsefly: *Tabanus atratus*   Fig. 105

Horseflies and deerflies both belong to the family Tabanidae, and most kinds suck the blood of larger mammals, including man. Some horseflies are, for flies, very large—the black horsefly can reach a length of 2.5 cm. Its whitish abdomen looks blue when it is flying, and it is the original "blue-tailed fly." Other kinds are generally smaller, some as little as the common housefly. The notorious yellow fly can make outdoor life a real burden, as its bite is painful and may lead to severe reactions in persons with allergies. Deerflies (*Chrysops*) of many kinds occur in Florida and can generally be distinguished from horseflies by their golden-spotted eyes and dark-banded wings. All are less than 1.5 cm long. Greenheads are another sort of horsefly with striped bodies and bright green eyes. They frequent coastal areas and can be a real pest along some of our

103. Treehopper
Entylia carinata

102. Cicada
Tibicen sp

104. Glassy-winged sharpshooter
Homalodisca coagulata

105. Black horsefly
Tabanus atratus

beaches, especially where these are near saltmarshes. All the kinds of Tabanidae that take blood as food inject an anticoagulant into the skin of an animal and then suck up the blood that collects as a small pool beneath the skin. In all these species of fly it is only the females that suck blood; males are innocuous pollen and nectar feeders.

From Carrabelle Beach on the northwest coast along a 200-mile stretch of beach to Pensacola Beach, there is an additional blood-sucking fly, *Stomoxys calcitrans* (Diptera: Muscidae), locally known as the dog fly, concentrations of which can be bad enough to drive bathers from the beaches. The flies breed in *Sargassum* seaweed washed onto the beach. They are most troublesome between the end of August and October and may penetrate inland up to 15 miles or offshore to bother fishermen (King and Lennert 1936).

Tachinid fly: *Archytas* sp.   Fig. 106

Tachinid flies are bristly in appearance and active in habits. Females lay their eggs on the larval bodies of other insects and their own larvae hatch and bore inward, living as internal parasites and eventually killing the host. Tachinids are thus important in the biological control of many insect pests.

Love bug: *Plecia nearctica*   Fig. 107

Love bugs have spread eastward from Mississippi and Louisiana since 1940 (Buschman 1976). Ask your grandfather about them, and he *should* agree that they have become a lot more numerous in recent years. This increase in numbers has led some to believe they have been released in recent times, intentionally or not, by the University of Florida during an experiment that somehow went awry. Wrong—they have been expanding their range at about 20 miles per year and reached Pensacola by 1949, Tallahassee by 1957, Gainesville by 1966, and South Florida by 1975.

These flies are most commonly encountered in coupling pairs (hence the name), either on the ground, in midair, or, most commonly, on car windshields. They are attracted to highways by a combination of ultra-violet light and automobile exhaust fumes, which must contain some active ingredient (Callahan and Denmark 1973). The decaying vegetation and mowed grass of roadside verges provide an ideal breeding place for the larvae. It is possible that the monthly roadside grass mowing by the Florida Department of Transportation, initiated about 1974, may have greatly exacerbated the lovebug explosions of the mid to late 1970s, because mowed grass was routinely left to rot in roadside ditches. Flies are seen during May and September, as there are two generations a year. The situation on highways is often so serious that love bugs become a pest, clogging radiator grills and smearing windshields. Because the insect is only active from around 10 A.M. onward, however, the early-morning motorist will miss most of the flies. They are also inactive at night. Love bugs should be washed off automobiles as soon as possible, for they may otherwise damage the paintwork. Adult bugs are quite harmless; they feed on nectar and pollen and do not bite or sting.

Known locally as "honeymoon" flies in certain parts of their range, copulating couples have been reported at altitudes up to 1,500 ft. by Florida Highway Patrol airplane pilots. Males live about 2–3 days, females a week or longer. The recorded flight periods of about a month are thus due to a continual replacement of flies by newly hatching individuals. The larger and stronger female controls flight and walking activity of the tandem pair, and copulation continues until the male dies (Hetrick 1970).

Bee hunter: *Laphria* sp.   Fig. 108

At first glance this fly looks like a bee, but in fact the bee hunter preys on bees, seizing them near the flowers they frequent. It is one of many examples in the fly world in which harmless species look like stinging Hymenoptera. Others include numerous hover flies, which mimic wasps. This phenomenon is known as Batesian mimicry after the famous naturalist H. W. Bates, who first drew attention to this phenomenon in 1861. The bee hunter is a member of the family Asilidae, the robber flies. Many other members of this family are common in Florida. All are very bristly, with a bearded face, and range from 1 to 2.5 cm long. Insect prey are generally taken on the wing, including insects as large as dragonflies.

Long-legged fly: *Condylostylus* sp.   Fig. 109

Long-legged flies, of the family Dolichopodidae, appear as tiny emerald jewels flitting around on vegetation. Most are small, less than 1 cm long, and have a beautiful metallic green or coppery sheen. The males, for their size, have very large genitalia, which are folded back underneath the body. Despite their small size, dolichopodids are predator on other insects, and their larvae feed on aquatic organisms.

106. Tachinid fly
    Archytas sp.

107. Love bug
    Plecia nearctica

108. Bee hunter
    Laphria sp.

109. Long-legged fly
    Condylostylus sp.

110. Paper wasp
Polistes sp.

Wasps, bees, and ants: order HYMENOPTERA

The most pertinent features of these insects to entomologists and laymen alike is their ability to sting, using a modified ovipositor. Fire ants, paper wasps, and bumble bees can all leave us with a powerful reminder of our hymenopteran encounters. However, the "good" done by members of the Hymenoptera far outweighs the harm. As pollinators, bees play a vital role in the pollination of much of the earth's vegetation. In North America alone the value of food crops pollinated by bees is at least $4.5 billion annually, and the yearly value of the world's honey crop is about $250 million. Furthermore, many species of wasps are parasitic on other insects and act as "biological control agents," keeping in check the hordes of insect pests. Ants in North America recycle as many soil nutrients as earthworms. Most Hymenoptera have biting and chewing mouthparts for collecting prey and, in the case of bees and some wasps, tonguelike structures modified for drinking nectar. The variety of lifestyles is astonishing, as hymenopterans are represented by over 100,000 species worldwide, second only to the beetles in diversity. Another interesting feature is that many species are social and live together in large nests (Tissot and Robinson 1954). The young are commonly legless grubs. Wasp nests in the wild are often constructed under tree limbs for protection from the elements, so homes, with their abundance of overhanging eaves, commonly provide ideal nesting places for wasps. The insects rarely sting unless disturbed.

111. Potter wasp nest
Eumenes fraternus

Paper wasp: *Polistes* sp.   Fig. 110

Paper wasp nests are common on eaves of buildings, ceilings of porches, and other outdoor structures. The nests are constructed of grey paperlike material formed of a mixture of chewed weather-worn wood and saliva. They are attached by a thin but strong pedicel often darkened by an ant-deterring coating. The wasps are not overly pugnacious but when molested can sting severely. The developing grubs are fed a mixture of chewed insect prey. The queen is similar in appearance to the workers. In hot weather the adults may be seen fanning the nest with their wings to maintain the correct temperature. Paper wasps belong to the family Vespidae, which includes the yellow jackets and hornets. Other members of the family include solitary species, some of which construct mud nests for their larvae.

Potter wasp nest: *Eumenes fraternus*   Fig. 111

This simple potlike nest is constructed by the potter wasp, *Eumenes fraternus*, a small black wasp with yellow markings. The female provides each mud pot with an anesthetized caterpillar or sawfly larva and seals the entrance with mud before moving on to fashion her next pot. The time taken to build a nest no doubt varies with the distance from the source of mud but is usually about 3–4 hours.

Thread-waisted wasp: *Eremnophila aureonotata*   Fig. 112

Thread-waisted wasps construct short burrows in the ground as their nests, generally in sandy places. The terminal chamber is enlarged and provisioned with stunned insect prey for the young larvae. Adults can often be seen feeding on nectar at flowers, where they may also catch their prey. The females use small physical landmarks in their environment, such as the positions of stones or pine cones, to remember the positions of their burrows, which they periodically revisit to stock them with more food for the larva. The interested observer can verify the orienteering ability of females by moving objects close to the nests, in much the same way as did the famous French entomologist Jean Henri Fabré in the 19th century when he first described their behavior.

Mud dauber nest   Fig. 113

Some species of sphecids construct long tubular mud cells, which they stock with immobilized prey, usually spiders. Nowadays such nests are commonly found underneath buildings or bridges. Females can often be seen at river banks, where they use their mandibles to shape mud into balls, which they carry back to the nest. The nests are often constructed side by side, giving the appearance of organ pipes and the common name to the wasp of "organ-pipe builder." The herring-bone appearance of some of these nests, especially those constructed by the genus *Trypoxylon*, results because successive loads of mud are spread on alternate sides. Although nests can readily be seen in human dwellings, they are less easy to locate in their natural surroundings under tree limbs or, as here, on vegetation. *Trypargilum politum* is the most common mud dauber nesting under highway bridges and farm buildings in Leon, Wakulla, Gadsden, and Liberty counties (Trexler 1985).

Hunting wasp: *Campsomeris quadrimaculata*   Fig. 114

Hunting wasps are large members of the Sphecidae. They construct tunnels in the ground, commonly along roadsides or under outside sheds. One or two insects, sometimes cicadas or beetle larvae, are placed in each cell at the end of the tunnel along with one egg. Cicadas squeak loudly when caught until they are paralyzed by the wasp's sting. All sphecid wasps can attack humans if interfered with, and some, such as the cicada killer, *Sphecius speciosus*, have particularly painful stings. Most, however, are not aggressive and will not sting even when handled.

Honey bee: *Apis mellifera* Fig. 115

Settlers brought the honey bee to North America in the 17th century; the species has been transported worldwide by man. There are many native species of Floridian bees, and wing venation is the commonly used taxonomic tool for classifying them. Honey bees have a complex social structure built around the queen, who lives for up to five years and produces up to 80,000 workers in her lifetime. The hive is constructed in a hollow tree or in beekeepers' hives. Although the food store is sufficient to maintain the colony through the winter, all males are evicted from the nest at this time. They quickly die, but more are produced by the queen the next spring. Adults are most commonly seen at flowers drinking nectar and collecting pollen, which they brush off and store in special "baskets" on their hind legs. Larvae are fed bee bread, a mixture of pollen and honey, the latter being predigested nectar stored in the cells of the hive.

One of the greatest concerns of Florida apiarists is the spread of killer bees, actually Africanized honey bees, from Central America to the Caribbean. In the 1950s, a breeding program was established in South America involving ordinary honey bees and aggressive African varieties. In 1957, twenty-six swarms of the latter escaped and became established in the wild. They have been expanding their range and mating with the European variety ever since. The aggressiveness of the African bees is genetically dominant and thus is passed on to the hybrid, which is more difficult to keep commercially. Killer bees are already present in Mexico, and predictions are that they should reach Florida by 1990.

American bumble bee: *Bombus pennsylvanicus* Fig. 116

This species of bumble bee creeps into Florida from its stronghold in the rest of the U.S.A. Like the honey bees, bumble bees also drink nectar and produce honey for their young to feed on but are not as socially advanced as their honey bee cousins. The colony dies back over winter, and only young queens survive to begin the underground nest anew in the spring.

Halticid bee: *Augochlora* sp. Fig. 117

Halticid bees have a shining metal-green body coloration and are a joy to watch flitting from flower to flower. Females dig nests in dead wood or use preexisting burrows and deposit pollen balls and nectar as a food supply for their larvae. Some halticid bees are also known as sweat bees for their habit of lapping up human or animal sweat. If grasped while engaged in this activity, they will sting.

Fire ant: *Solenopsis invicta* Fig. 118

Both the red imported fire ant, S. *invicta*, and the black imported fire ant, S. *richteri*, were introduced from South America into the United States at Mobile, Alabama—the former about 1940, the latter about 1918. *Solenopsis richteri* is more or less confined to Alabama and Mississippi, but S. *invicta* has become widespread from Texas to the Carolinas, including Florida. Its northward spread has now slowed down. Although many exotic insect pests have been inadvertently introduced into Florida, few have had the dramatic impact of the fire ant (Lofgren 1983). Fire ants sting and bite viciously, producing a burning sensation; hence the common name. Nests are commonly encountered on open ground, and large amounts of soil are excavated. Larger mounds may contain more than 100,000 ants. In heavily infested areas, there may be more than 30 nests per hectare. Many verdant pastures have become happy construction sites for colonies of fire ants. This photograph shows fire ants within their nest, their black abdomens reflecting circles of light from the camera flash.

Cow killer: *Dasymutilla occidentalis* Fig. 119

This antlike insect is in fact a flightless wasp, family Mutillidae. Most mutillids are brightly colored and are encountered walking over open ground. Their densely hairy bodies make them appear as if they are clothed in velvet. The sting of this insect is so severe that some people claim it could kill a cow. The female in her wanderings is actually searching for a bumble bee nest to oviposit in, where the hatching larvae will be able to feed on bee grubs. The smaller males are winged and can fly.

112. Thread-waisted wasp
Eremnophila aureonotata

113. Mud dauber nest

114. Hunting wasp
Campsomeris quadrimaculata

115. Honey bee
Apis mellifera

116. American bumble bee
Bombus pennsylvanicus

117. Halticid bee
Augochlora sp.

118. Fire ant
Solenopsis invicta

119. Cow killer
Dasymutilla occidentalis

Wasps, Bees, and Ants: Order Hymenoptera ⌣ 71

Beetles: order COLEOPTERA

This is the largest order of living things, with about 300,000 described species, 27,000 in North America. Every fifth living thing is a beetle. No other organisms come close to their staggering diversity in terms of number of species. There are as many kinds of beetles as there are plant species in the world. Indeed many beetles are herbivorous, and few plants are without beetle damage, either feeding scars on the leaves or larvae in the trunk and roots. Beetles range in size from the largest insects, 120 mm long (the Goliath beetles of Africa), to the smallest at 0.025 mm long. With their tough armored coating, they are the tanks of the insect world, commonly found trundling along the ground. Yet most beetles can fly as well, and many are attracted to lights at night. Metamorphosis in these insects is complete; that is, the larvae do not resemble adults but are grublike creatures that transform themselves at the pupal stage.

Patent leather beetle: *Odontotaenius disjunctus* — Fig. 120

Patent leather beetles, or bessbugs as they are also known, are among the most common inhabitants of rotting logs. These large shiny black beetles and their grublike larvae live in loosely organized colonies in tree stumps. Both larvae and adult are quite vocal and can produce squeaking noises by rubbing various parts of their anatomy together.

Unicorn beetle: *Dynastes tityus* — Fig. 121

The unicorn beetle or Eastern hercules beetle is the largest beetle to be found in Florida, males reaching 6.5 cm in length. The male bears a single arching horn on both head and pronotum, but the female (illustrated here) has none. Despite its impressive appearance, the unicorn beetle is quite harmless. The young are large white grubs that feed on plant roots in the soil. The genus *Dynastes* includes one of the world's largest beetles, D. *hercules*, which has a huge pair of horns, up to 100 mm in length, and is common in Venezuela and the lower Antillean islands.

Ox beetle: *Strategus antaeus* — Fig. 122 & 123

Ox beetles, sometimes also known as rhinoceros beetles, are brown to black in color, and the males have three distinct horns on the pronotum. Like the unicorn beetles, ox beetles are members of the family Scarabaeidae, the scarabs. There are an immense number of scarab beetles in the United States—at least 1,375, with probably more to be described. Many live in the soil on roots, some live in rotten wood or carrion, and some in animal dung. The latter are familiar as dung-ball rollers, burying the balls and thus providing a food supply for their upcoming larvae, which hatch from eggs laid on the dung. *Phanaeus igneus*, often seen on the ground in northern Florida in the fall, is unmistakable, with green and red iridescence (White 1983).

Green June beetle: *Cotinus nitida* — Fig. 124

The green June beetle is a hornless scarab beetle. Its larvae feed on roots and humus in lawns and gardens. Whereas many scarabs are often caught at night flying around lights, green June beetles are active in the daytime. In high summer, the loud buzzing of flying beetles may be apparent around trees growing from abundant turf, as males and females congregate, mate, and drop to the grass to oviposit (Patton 1956).

Spotted pelidnota: *Pelidnota punctata* — Fig. 125

The spotted pelidnota or grapevine beetle feeds on leaves and fruits of wild and cultivated grapes. Larvae eat decaying wood in tree stumps. The key for the recognition of this species is the row of three black dots on the side of each elytron (wing case).

Palm weevil: *Rhynchophorus cruentatus*   Fig. 126

The palm weevil is the largest of Florida's weevils, measuring up to 3.5 centimeters long. It feeds in the trunks of palms such as cabbage palmetto and royal palm. The adults are attracted to fermenting sap and damaged palms. Here they lay eggs, and the large grubs develop inside the trunk. The palm weevil is a member of the Curculionidae, the most species-rich family on earth with at least 2,500 species in America. Most members of the family are much smaller, but all exhibit the characteristic elongate snout with mouthparts at the end, hence their alternative name of snout beetle. Exit holes of other weevils can commonly be found in the fallen acorns of oak trees.

Tortoise beetle: *Hemisphaerota cyanea*   Fig. 127

This species, with its brilliant metallic blue coloration, is commonly found feeding on the leaves of palmetto plants in Florida and Georgia. In general chrysomelid beetles are often brilliantly colored. The tortoise beetles are a subset of the family Chrysomelidae, the leaf beetles, small beetles similar to ladybird beetles in appearance but whose adults feed generally on herbaceous foliage. The tortoiselike shape of this beetle and of its close relatives lends itself to the popular name. The larvae, short and grublike, have forked posterior appendages bent forward over the body, holding a wad of cast skins and excreta. If a larva is threatened by an enemy, perhaps an ant, this collection is thrust into the predator's face.

Carolina sawyer: *Monochamus carolinensis*   Fig. 128

The long antennae of the Carolina sawyer help identify it as a member of the family Cerambycidae, the long-horned beetles, the sixth largest family of beetles in the United States, with over 1,100 widely distributed species. All are plant feeders whose larvae (often called round-headed borers) eat the solid tissues (stems, roots, and trunks) of trees. The characteristic feature of long-horns is their antennae, which are at least half as long as the body and sometimes much longer. Handle long-horns carefully; they bite.

Ivory-spotted borer: *Eburia quadrigeminata*   Fig. 129

Adults of this long-horned beetle can be easily recognized by their two pairs of ivory-colored swellings on each elytron. Adults measure about 14–24 mm in length and have a brownish ground color. Larvae feed on inner bark and sapwood.

Cylindrical hardwood borer: *Neoclytus acuminatus*   Fig. 130

At first glance, adult *Neoclytus* beetles are easily mistaken for wasps. They act like them, too, being quick to take off from flowers if approached, not at all cumbersome beetlelike behavior. Whether or not they are wasp mimics, the larvae are still wood feeders and can honeycomb sapwood, causing serious damage to logs left in the woods or stored with their bark on. Adults emerge in the first warm days of spring.

Blind click beetle: *Alaus myops*   Fig. 131

Click beetles (family Elateridae) are known best for their ability to click. If placed on its back, a click beetle will bend its head and prothorax back and then snap its body straight, producing an audible click and flipping itself into the air. Individuals will repeat this performance until they right themselves. The blind click beetle is about 25–40 mm long and, despite its name, possesses distinctive eyespots. However, the even larger eyed click beetle, A. *oculatus*, found in locations further north and east, has even more prominent distinctive eyespots. Like many other click beetles, the larvae are found in rotten logs; other species live in soil. Collectively, the larvae are known as wireworms for their distinctive long, narrow, and hard bodies.

Tiger beetle: *Cicindela scutellaris*   Fig. 132

Tiger beetles (family Cicindelidae) are about 1.5 cm long and often brightly colored. The common name refers to the predaceous habits of the adults, which feed on other insects. Most are found in open, sandy areas such as beaches, paths, and trails. They run fast and fly readily, especially at the approach of an entomologist. The larvae are predaceous, constructing vertical burrows and ambushing passing insects.

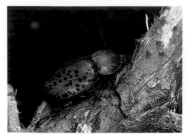

121. Unicorn beetle
Dynastes tityus

122. Ox beetle
Strategus antaeus

123. Larva of ox beetle
Strategus antaeus

124. Green June beetle
Cotinus nitida

125. Spotted pelidnota
Pelidnota punctata

120. Patent leather beetle
Odontotaenius disjunctus

127. Tortoise beetle
Hemisphaerota cyanea

126. Palm weevil
Rhynchophorus cruentatus

128. Carolina sawyer
Monochamus carolinensis

129. Ivory-spotted borer
Eburia quadrigeminata

130. Cylindrical hardwood borer
Neoclytus acuminatus

131. Blind click beetle
Alaus myops

132. Tiger beetle
Cicindela scutellaris

133. Fiery searcher
Calosoma scrutator

134. Firefly
Photinus sp.

Fiery searcher: *Calosoma scrutator*   Fig. 133

Also known as the caterpillar hunter, this insect is highly beneficial because it attacks caterpillars on shrubs and trees. The family in general (Carabidae) are known as ground beetles, and they can most often be found wandering on the ground. They are nocturnal, hunting at night and hiding under debris or bark during the day. This is one of the most spectacular carabids; others have a general black body color, though sometimes with a metallic sheen. Carabids, especially this species, omit a very pungent odor when disturbed.

Firefly: *Photinus* sp.   Fig. 134

Fireflies or lightning bugs are commonly seen in Florida on summer evenings as they blink their green or yellow light organs to attract mates. Fireflies are not in fact flies at all but belong to the beetle family, Lampyridae. There are many species, each with a characteristic flashing pattern, specific to the species. The two main genera are *Photuris* and *Photinus*. Some species have even evolved to mimic the flash of other species, which they attract and then eat (termed aggressive mimicry). The chemical process that creates the light is nearly 100% efficient—almost no energy is given off as heat; compare this with an electric light bulb! The light comes from light-producing organs on the underside of the abdomen. These appear ivory-colored during the day but even then the beetle can be induced by a gentle squeeze to produce flashes.

There are also to be found in Florida three species of luminous elaterid beetles, *Pyrophorus atlanticus*, *P. havaniensis*, and *P. phosderus* (Moznette 1920). The flash of the elaterids is more constant when the insects are in flight, unlike those of the lampyrids, which are intermittent. The situation of the light-producing organs in these species is also different, being on the pronotum or thorax instead of on the underside of the abdomen.

Miscellaneous Insect Families

The following insects, although commonly encountered in Florida, do not conveniently fall into any one taxonomic category or are not easily rec- ognized as doing so. They are dealt with here in a separate section.

Palmetto stick insect: *Anisomorpha buprestoides*   Fig. 135

Stick insects belong to the order Phasmida, large insects that closely resemble twigs and sometimes leaves. They are foliage feeders on shrubs, tall herbs, and trees and are most active at night. Their resemblance to dead twigs and leaves is uncanny, and such camouflage offers great protection from visually hunting predators. *Anisomorpha buprestoides* is often seen on palmetto in Florida and will commonly crawl inside a box left overnight in a palmetto stand. Frequently, one can find mating pairs, the smaller male riding on the female's back. Matings have been known to last three weeks (Clark 1974)!

The other large stick insect sometimes found in the northern part of the state is *Diapheromera femorata*, the northern walking stick, even more elongated and sticklike. Males are brown; females and juveniles greenbrown. When disturbed, stick insects exude a chemical containing alkaloids that can cause skin burns.

Praying mantis: *Stagmomantis carolina*   Fig. 136

Praying mantids are present in the warmer regions of the world and are especially well represented in the tropics. They are day-time carnivores specialized to feed on other insects. Commonly camouflaged to match the plants they rest on, these insects are often seen perched on low herbage with gin-trap forelegs poised ready to strike. If not treated with caution, an adult mantid will strike at an observer and, if picked up, can bring its mobile forelegs backwards over the body to snare the fingers. If these distractions fail, mantids will sacrifice a limb or two in order to escape, a process known as autotomy. Front legs are rarely autotomized, however, because the mantid would surely starve without them. The powerful jaws of mantids can easily slice through other insects' armor plating, and those of the female regularly bear down upon the unfortunate male during or after the courtship act (as shown here); the male is a valuable source of protein for her developing eggs. *Mantis religiosa*, perhaps the "original" praying mantis, was originally native to Europe and was introduced accidentally into the U.S. in 1899 on European nursery stock, probably as eggs. Its wing tips, unlike those of *Stagmomantis*, do not extend beyond its abdomen, and it does not occur in Florida.

American cockroach: *Periplaneta americana* Fig. 137

The American cockroach, "waterbug," or "palmetto bug" is an inhabitant of the crannies and crevices of tropical and subtropical vegetation and of the insides of warm buildings. The adults fly well and are attracted to artificial lights at night, whence they often enter dwellings. These insects are true omnivores, being able to eat almost anything. Most active at night, cockroaches are very sensitive to vibrations and air movements, which they sense with the two appendages (cerci) that project from the end of the abdomen. Thus, cockroaches often scuttle away long before an attacker is close enough to strike.

Other cockroaches found in Florida are the German cockroach, *Blatella germanica*, light brown with two longitudinal stripes on the pronotum, and, since 1985, the Asian roach. Asian roaches are much more of a pest than their close relative, the German roach, from which they cannot easily be distinguished. Asian roaches will fly toward well-lit areas, including illuminated walls and television screens, making them quite a nuisance (Koehler and Patterson 1987). Known from only a six-square-mile area north of Lakeland in 1986, these roaches have spread and can now be found from Tampa–St. Petersburg through Lakeland. They were probably originally introduced via the port of Tampa on some Asian consignment.

Gall produced by gall insect Fig. 138

The following types of insects are known better for the damage they produce on plants than from the adults themselves.

Gall insects include insignificant tiny wasps and minute true flies. Regardless of the group, the adult insect lays its eggs in hosts plants, in either the stems, twigs, or leaves, and the young insect begins to grow. As it grows, it induces the host tissue to grow over it, thus forming a protective layer and one that is rich in protein, which provides the young larva with a personal source of food. Each species of gall fly or wasp, of which there are many, produces a characteristic type of gall, characteristic in shape or size or in its position on the host plant. Each species of gall insect is much more recognizable by its gall than by the adult itself. Oak trees are particularly hard hit by gall makers, and early botanical drawings show galls as representative of the normal plant. No fewer than 40 kinds of galls are produced on roses by gall wasps.

Leaf mine produced by leaf miner: *Tischeria citripennella* Fig. 139

Many trees, shrubs, and herbs throughout Florida are infested with leaf miners (Stiling 1988a). Collectively, this type of feeding method is employed by larvae of tiny moths, flies, wasps, and beetles, each of which usually attacks a particular kind of plant (Needham et al. 1928). The delicate adults lay their eggs on leaves, and the larvae burrow between the leaf surfaces, eating all the tissue in between and creating characteristic leaf mines that appear blisterlike to the observer (Hering 1951). As with gall insects, it is the larval damage that is most characteristic; the adults themselves are difficult to distinguish. Some miners produce large blotch-shaped mines, whereas others create long meandering serpentine tunnels. The mine illustrated is on water oak, *Quercus nigra*, and is caused by a moth. Other common miners on oaks in Florida include *Stilbosis quadricustatella* (Lepidoptera, blotch mine with frass on leaf underside), *Cameraria* sp. (Lepidoptera, large blotch mine), *Brachys ovatus* (Coleoptera, mines virtually a whole leaf of *Quercus virginiana*, shiny black egg case also usually visible), *Stigmella* (Lepidoptera, a serpentine mine), and *Acrocercops* (Lepidoptera, a thin, right-angled mine). For fuller accounts, see Connor et al. (1983), Faeth et al. (1981), Stiling et al. (1987).

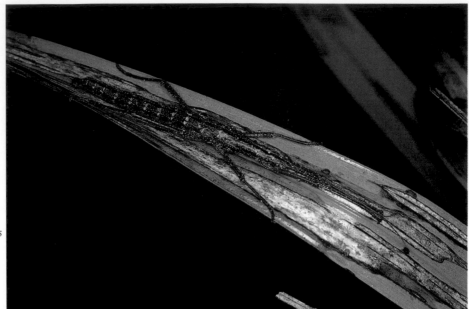

135. Palmetto stick insect
Anisomorpha buprestoides

136. Praying mantis
Stagmomantis carolina

137. American cockroach
Periplaneta americana

139. Leaf mine produced by leaf miner
Tischeria citripennella

138. Gall produced by gall insect

140. Hentz's striped scorpion
Centruroides hentzi

141. Daddy-long-legs
Leiobunum sp.

142. Tick
Dermacentor sp.

143. Golden silk spider
Nephila clavipes

144. Black and yellow argiope, adult
Argiope aurantia

145. Black and yellow argiope, juvenile
Argiope aurantia

Spiders and relatives: class ARACHNIDA

Most people think of insects and spiders in the same terms: creepy crawlies. In fact, spiders and their kin are placed in a different class, the Arachnida, characteristic features of which include the possession of eight legs and a body consisting of two segments, the cephalothorax (or prosoma) and the abdomen (or opisthosoma). Unlike insects, arachnids do not possess wings or antennae. The group is diverse and includes spiders, scorpions, harvestmen, ticks, and other less commonly encountered orders. Such animals are among the least loved on earth. Some are poisonous to humans (spiders and scorpions) or parasitic on them (ticks). For the most part, arachnids are predators on other arthropods, especially insects and occasionally other arachnids. Harvestmen may also eat some decaying plant material, whereas many mites are also plant feeders or are parasitic on other animals. Ticks are exclusively parasitic on vertebrates, including man. The class Arachnida is an ancient one; early fossils of scorpions date back to the Silurian period some 400 million years ago. Most species take their food in liquid form, pumping salivary juices into the prey and sucking out the resultant soup.

146. Silver argiope
Argiope argentata

147. Arrow-shaped micrathena
Micrathena sagittata

148. Crablike spiny orb weaver
Gasteracantha cancriformis

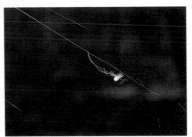

149. Orchard spider
Leucauge venusta

150. Green lynx spider
Peucetia viridans

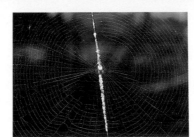

152. Cyclosa spider
Cyclosa turbinata

151. Carolina wolf spider
Lycosa carolinensis

Hentz's striped scorpion: *Centruroides hentzi* Fig. 140

None of the three species of scorpion commonly found in Florida is capable of inflicting a lethal sting, though the site of the sting may be slightly painful for several hours. The venom is a neurotoxin, comparable in mode of action to that of a cobra snake, but the dose is insufficient to kill a human. In the United States, only species in the arid southwestern deserts are dangerous to people. Most scorpions in the state of Florida are dull yellow with brown markings and about 2.5 cm long, though *C. gracilis*, reddish brown to nearly black, can be as long as 15 centimeters (Muma 1967). All are nocturnal and predaceous, coming out from their day-time resting places under logs and boards to hunt insects and spiders. The lobster-claw-like pedipalps or pincers hold the prey while it is stung to death. Females give birth to live young, resembling tiny adults, which ride on the back of the mother until they molt for the first time. Juveniles grow slowly, some taking up to five years to become adults.

Daddy-long-legs: *Leiobunum* sp. Fig. 141

Daddy-long-legs are not true spiders but belong to the closely related order Opiliones. Most species have very long legs, but the taxonomically distinguishing feature is the broad joint of cephalothorax and abdomen. In true spiders, these two body parts are joined by a narrow waist or pedicle. Another unusual feature is that a male opilionid has a penis, which it inserts into females to effect fertilization. Other members of the Arachnida have simple genital openings. Opiliones are also commonly called harvestmen, the first species to be described having been collected at harvest time. The long legs of the daddy-long-legs are fragile and often break off, but they cannot be regenerated.

Tick: *Dermacentor* sp. Fig. 142

Dermacentor ticks are members of the famiy Ixodidae or hard ticks, so named because of a hard plate on top of the body. This plate makes it very hard to crush ticks between thumb and forefinger. The adults are small, about 3 mm long and feed on larger animals, especially deer. They usually cling to wayside plants beside trails, waiting to transfer to passing hosts. After sucking out a blood meal, the tick drops to the ground to molt. This process is repeated until the parasite is mature. Unfortunately, humans are often included as possible hosts, and, though the ticks can easily be picked off, many diseases (such as Rocky Mountain spotted fever) can be transmitted between hosts in this way. Fortunately, very few such cases have been reported in Florida. Among the most frightening are the stories of paralysis in young children induced by attachment of the American dog tick, *Dermacentor variabilis* (Heidt 1954). The paralysis starts at the legs and works it way up the body. It is caused by toxin secreted from the salivary glands of the pregnant engorged ticks. The removal of the tick brings about rapid recovery, although sometimes ticks are hard to find when they lodge in the scalp.

Florida woods are also home to closely related mites of the family Trombiculidae, the dreaded chiggers or red bugs. There are five species that may be troublesome to humans in north-central Florida and several that bother animals, including one that can cause severe losses to the turkey industry (Rohani and Cromroy 1979). Chiggers are mostly lymph-feeding ectoparasites of vertebrates. They commonly concentrate their attacks on areas of skin tightly wrapped by clothing such as that beneath waistbands or under socks. The resultant red itching area of skin is caused by an allergic reaction to chigger saliva. Irritating though this reaction may be (for up to several days), in other parts of the world chiggers transmit severe diseases such as Tsutsugamushi fever (or scrub-typhus). Chigger bites can be reduced in number by vigorous washing after walks in the woods, or they can be prevented entirely by use of creams or powder containing sulphur, which are available for the purpose.

Golden silk spider: *Nephila clavipes* Fig. 143

The golden silk spider, with its tufted legs and large body, is characteristic of Florida and the lower south-eastern United States. It is also found throughout the neotropics, the tropical areas of the New World, down to Brazil. This spider is one of the largest Floridian members of the family Araneidae, the orb weavers, and it builds a particularly strong web whose silk appears as thick as horse hair. Walking into one of these large webs, which may measure up to a meter across, is quite a shock, although the spider usually scuttles quickly away. It is the large females that are most commonly seen; males are less than 1 cm in body length.

Unlike many other orb weavers, *Nephila* does not attack insects by initially wrapping them in silk. Instead it bites its victims, injecting them with venom. Because it cannot use silk from a long distance and because it must come into fairly close contact with the prey in order to bite it, *Nephila* is itself in some danger of being bitten or kicked by the prey. As a result, and in spite of its size, *Nephila* is a "cowardly" spider and cannot deal effectively with large or aggressive insects (Lubin 1983).

Black and yellow argiope: *Argiope aurantia* — Fig. 144 & 145

This spider, and all the spiders illustrated here except the green lynx and Carolina wolf spiders, are members of the family Araneidae, the orb weavers, a huge family with several hundred North American species. They spin spiraling orb webs on support lines that radiate out from the center. The females have no nest and remain always in the center of the web. The smaller males often build their webs in outlying portions of the females' webs. Juveniles have distinct black and white striped legs.

Silver argiope: *Argiope argentata* — Fig. 146

The web of the silver argiope contains zigzag cross strands, which form a characteristic cross-shaped structure at the center. This device, known as a stabilimentum (see Figure 145), is a warning for birds that a web lies in their flight path (Eisner and Nowicki 1983). The logic is that birds will avoid the web, thus sparing it for use another time. As in all species of orb weavers, females rest head down at the center of the orb. This species of spider is not particularly tolerant of frost and is therefore more associated with tropical areas of the New World.

Arrow-shaped micrathena: *Micrathena sagittata* — Fig. 147

This is a spiny spider, the female having two dark-red diverging spines on the abdomen and two pairs of shorter tubercles, red with black tips. The web is spun in a vertical plane and contains a central hole across which the spider rests.

Crablike spiny orbweaver: *Gasteracantha cancriformis* — Fig. 148

A small orb weaver with females only about 8–10 mm long, the crablike spiny orb weaver builds a vertical orb with a few spiral strands at the center. It is also visible on single strands of silk spun between trees. As with most orb weavers, the web is replaced daily, a new one being spun up in the early evening in about an hour.

Orchard spider: *Leucauge venusta* — Fig. 149

The orchard spider is a member of the long-jawed orb weavers, or Tetragnathidae, characterized by relatively large, powerful jaws. It is very small, however, measuring only 4–7 mm, and its jaws pose no threat to human beings. These spiders hang on the underside of the web or rest on a nearby stem waiting for their prey to blunder into the larder. Other tetragnathids spin no webs but sit motionless among grass stems with two pairs of legs stretched out straight in front.

Green lynx spider: *Peucetia viridans* — Fig. 150

Among the non-web-building spiders, the green lynx spiders are probably the most commonly seen. They wait on grasses, flower heads, and low shrubs for prey, which they jump onto with very quick movements. The lynx spiders, family Oxyopidae, are mainly tropical spiders, and the green lynx is found in the southern United States and Mexico.

Carolina wolf spider: *Lycosa carolinensis* — Fig. 151

Among the ground-dwelling spiders, the wolf spiders (family Lycosidae) are the largest and fastest, chasing and dragging down their insect prey. Among the wolf spiders, the Carolina wolf spider is the largest and most commonly seen in Florida. It is one of the biggest spiders in the entire United States. Females make a hole, often in sand or in the base of a tree stump, and hide in it with their eggs; they may also spin a silken cover for the nest hole.

Cyclosa spider: *Cyclosa turbinata* — Fig. 152

The first thing one notices about this spider is the web itself, with its line of rubbish neatly laid out across the center. This line consists of parts of dead insects and other material fastened in silk. In the middle of summer, egg cocoons are located there too. The spider sits in the middle of this band looking like part of the rubbish. When an old web is torn down, the band of rubbish is left in place and the new web built across it.

153. Millipede
Narceus sp.

Millipedes and Centipedes

Neither millipedes nor centipedes are true insects or spiders, but both are united with the latter under the banner of the phylum Arthropoda, a phylum that also contains the marine crustaceans and whose members all share the characteristics of a hard, chitinous exoskeleton and paired segmented appendages.

Millipede: *Narceus* sp.   Fig. 153

The genus *Narceus* contains some of the largest North American millipedes, which may reach 100 mm in length. Commonly, other Floridian millipedes are smaller, but all are wormlike in appearance, and each body segment bears two pairs of legs. Millipedes are generally found in damp places on the forest floor—under leaves, stones, or logs, where they scavenge on decaying plant material. Be careful in picking up millipedes because some can emit ill-smelling caustic fluids, and one entomologist I know was caused severe discomfort when this discharge was sprayed into his eyes. Hydrogen cyanide has been shown to be the active ingredient in certain sprays. More commonly, millipedes curl up and play dead when disturbed.

Centipede: *Scolopendra* sp.   Fig. 154

The centipedes are also wormlike animals but contrast with millipedes in the number of legs present—centipedes have only one pair per body segment—and in their method of feeding—centipedes are predaceous. Their general foods are other insects and some spiders, which they encounter while foraging on the ground. *Scolopendra* possesses poison jaws, which are used to paralyze prey. Although the smaller centipedes are quite harmless to man, larger individuals can inflict a painful bite, and in the Florida Keys very large *Scolopendra* are present and should be treated with some respect.

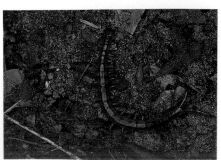

154 Centipede
Scolopendra sp

Creating a Butterfly Garden: Attracting Lepidoptera

Butterflies come into a yard for two reasons: to feed as adults and to lay eggs. Providing adult food is the easiest thing to do, but providing larval food, for the caterpillars, means that you can watch the entire life cycle of a butterfly and should ensure a plentiful supply of adults as well.

There are many flowers that are attractive to butterflies as nectar sources. A list of these is provided in Table 1. Some of the best include *Buddleia* sp., the so-called butterfly bush, and butterfly weed or milkweed, *Asclepias tuberosa* (Habeck 1984) There are more than 50 species of *Buddleia*, and about eight or nine have been cultivated in Florida at one time or another. The most widely grown is *Buddleia davidii*, which will grow up to 20 feet tall. The varieties of this plant have purple, lilac, pink, red, or white flowers. Plants can be propagated by seed or cuttings but grow spindly unless pruned.

Butterfly weed has yellow to orange-red flowers, which are attractive to many species of butterfly. Transplanting individuals from the wild is tricky, and they are best purchased from commercial nurseries. Of the other species listed, it is worth remembering that *Lantana* is poisonous, especially the berries, which are highly toxic to children. This feature is unfortunate, because it is one of the best natural butterfly plants in Florida. *Bidens* has sticky seeds and thistles, is prickly, and may be an inconvenience.

It will probably be difficult to establish all of the flowers listed in Table 1 in your garden, but a good rule of thumb is that sweet, pungent, and acid-smelling flowers attract butterflies. Particularly attractive colors include orange, yellow, pink, purple, and red. White flowers and those emitting fragrance at night tend to attract moths. Plants with deep-throated, drooping, or enclosed flowers are not well suited for nectar gathering. Wildflowers are great attractors; many hybridized flowers are not. Thus single marigolds are good butterfly plants, but hybrid varieties are not.

Providing host plants for butterfly caterpillars is a little more difficult and a little more unpopular because the plants involved are sometimes considered weeds. Furthermore, only 9% of the butterflies in the eastern United States are polyphagous, that is, are able to utilize a wide variety of plants. Most are monophagous or highly specific, using only one kind of plant or a very narrow range (Opler and Krizek 1984). Two "sweepstakes" plants that might be considered are *Cassia* spp. and *Passiflora* spp. (Habeck 1984). Passion flowers, *Passiflora*, can provide a nice drink and are host to three attractive butterflies, *Agraulis vanillae*, the Gulf fritillary, found throughout Florida; *Heliconius charitonius*, the zebra longwing, from Tampa southward; and *Dryas iulia*, the flambeau, limited to Dade and Monroe counties. *Cassia* spp. are host plants for several species of sulphur butterflies. A more complete listing of Florida butterflies and their host plants is provided for the enthusiast in Table 2.

The list of plants for moths would certainly be too long to print in this book, but there is another way to attract them—apart from the use of lights, that is. Most butterflies and moths don't just drink from flowers; they also absorb moisture and nutrients from moist sand or dirt. One can often see sulphur butterflies doing so en masse at a particularly wet patch, a phenomenon known as mud puddling. I have even seen swallowtails drinking from road kills! By making a sweet sugary attractant and spreading it on old logs, fence posts, and rocks, one can attract moths and butterflies, which will come to drink. This process is known as "sugaring." One can even soak sponges in a sugar solution and

hang them at dusk from trees. Return periodically
during the night to check your catch. All old timers
have their special sugar recipes, which they favor,
but a good basic one is as follows:

       one pound of sugar

       one mashed, overripe banana (or other fruit)

       one cup of molasses or syrup

       one cup of fruit juice

Blend these together, leave in the sun for an hour or
two to brew (no longer, or it will dry up), and paint
on in the late afternoon.

Table 1. Florida flowers attractive to butterflies. For time of flowering: sp = spring, su = summer, fa = fall,
yr = year round; for natural(ized) geographic location: N = north Florida, C = central Florida, S =
south Florida (from personal observations and after Hannahs 1986 and Habeck 1984).

Common name	Scientific name	Native	Location	Flowering Period
TREES				
Bay cedar	Suriana maritima	yes	C,S	yr
Bottlebrush tree	Melaleuca quinquenervia	no	C,S	yr
Chinaberry	Melia azedarach	no	N,C,S	sp
Citrus	Citrus spp.	no	C,S	sp
Hawthorn	Crataegus uniflora	yes	N	sp
Poisonwood	Metopium toxiferum	yes	S	sp
SHRUBS				
Butterfly bush	Buddleia lindleyana	no	N	su,fa
Fetterbush	Lyonia lucida	yes	N,C,S	sp
New Jersey tea	Ceanothus americanus	yes	N,C	sp
Red buckeye	Aesculus pavia	yes	N,C	sp
Sassafras	Sassafras albidum	yes	N,C	sp
HERBS, VINES, AND GROUND COVER				
Annual phlox	Phlox drummondii	no	N,C,S	sp
Blazing star	Liatris tenuifolia	yes	N,C,S	su,fa
Butterfly weed	Asclepias tuberosa	yes	N,C,S	su,fa
Clover	Trifolium spp.	yes	N,C	sp
Dotted horsemint	Monarda punctata	yes	N,C	su,fa
Goldenrod	Solidago fistulosa	yes	N,C,S	fa
Honeysuckle	Lonicera sempervirens	yes	N,C,S	sp,su
Ironweed	Vernonia angustifolia	yes	N,C	su,fa
Lantana	Lantana camara	yes	N,C,S	sp,su,fa
Pickerelweed	Pontedaria cordata	yes	N,C,S	sp,su,fa
Redroot	Lachnanthes caroliniana	yes	N,C,S	su
Spanish needles	Bidens pilosa	yes	N,C,S	sp,su,fa
Standing cypress	Ipomopsis rubra	yes	N,C	su
Swamp coreopsis	Coreopsis nudata	yes	N	sp,su
Thistles	Cirsium sp.	yes	N,C,S	sp,su,fa
Tickseed	Coreopsis gladiata	yes	N,C,S	su,fa
Verbena	Verbena brasiliensis	no	N,C,S	sp,su,fa

Table 2. Larval food plants for butterflies (after Habeck 1984, Hannahs 1986, Pyle 1981, and personal observations).

Butterfly	Larval food plants
Atala (*Eumaeus atala florida*)	coontie (*Zamia floridana*)
Buckeye (*Junonia coenia*)	plantain (*Plantago* spp.), snapdragon (*Antirrhinum* spp.), *Ludwigia* spp.
Caribbean buckeye (*Junonia evarete*)	black mangrove (*Avicennia*), blue vervain (*Stachytarpheta* sp.)
Cloudless sulphur (*Phoebis sennae*)	sennas (*Cassia* spp.)
Eastern black swallowtail (*Papilio polyxenes*)	Queen Anne's lace (*Daucus carota*), carrots, parsley
Florida purplewing (*Eunica tatila*)	unknown
Florida white (*Glutophrissa drusilla*)	capers (*Capparis*)
Giant swallowtail (*Heraclides cresphontes*)	citrus (*Citrus* spp.)
Gold rim (*Battus polydamus*)	pipevine (*Aristolochia* sp.)
Great southern white (*Ascia monuste*)	peppergrass (*Lepidium virginicum*), saltwort (*Batis maritima*)
Gulf fritillary (*Agraulis vanillae*)	passion flower (*Passiflora incarnata*)
Julia (*Dryas iulia*)	passion flower (*Passiflora incarnata*)
Malachite (*Siproeta stelenes*)	*Blechum* sp.
Miami blue (*Hemiargus thomasi*)	legumes, cat claw (*Pithecellobium keyense*)
Monarch (*Danaus plexippus*)	milkweeds (*Asclepias* spp.)
Orange-barred sulphur (*Phoebis philea*)	sennas (*Cassia* spp.)
Orange sulphur (*Colias eurytheme*)	white clover (*Trifolium repens*)
Painted lady (*Vanessa cardui*)	thistles (*Cirsium* spp.)
Palamedes swallowtail (*Pterourus palamedes*)	red bay (*Persea borbonia*), sassafras (*Sassafras albidum*), sweetbay (*Magnolia virginiana*)
Pearly crescentspot (*Phycoides tharos*)	asters (*Aster* spp.)
Pipevine swallowtail (*Battus philenor*)	pipevine (*Aristolochia* sp.)
Queen (*Danaus gilippus*)	milkweeds (*Asclepias* spp.)

cont.

Table 2. Larval food plants for butterflies (after Habeck 1984, Hannahs 1986, Pyle 1981, and personal observations).

Butterfly	Larval food plants
Question mark (*Polygonia interrogationis*)	hackberries (*Celtis*), nettles
Red-spotted purple (*Basilarchia astyanax*)	hawthorns (*Crataegus*), scrub oaks (*Quercus* spp.)
Ruddy daggerwing (*Marpesia petreus*)	figs (*Ficus* spp.), cashew (*Anacardium occidentale*)
Schaus' swallowtail (*Heraclides aristodemus ponceanus*)	torchwood (*Amyris elemifera*)
Spicebush swallowtail (*Pterourus troilus*)	spicebush (*Lindera benzoin*), various bays (*Persea* spp.), sweetbay (*Magnolia virginiana*)
Tiger swallowtail (*Pterourus glaucus*)	sweetbay (*Magnolia virginiana*)
Variegated fritillary (*Euptoieta claudia*)	passion flowers (*Passiflora* spp.), pansies
Viceroy (*Basilarchia archippus*)	plums (*Prunus* spp.)
White peacock (*Anartia jatrophae*)	ruellia (*Ruellia occidentalis*), water hyssop (*Bacopa monniera*)
Zebra longwing (*Heliconius charitonius*)	passion flowers (*Passiflora* spp.)
Zebra swallowtail (*Eurytides marcellus*)	pawpaw (*Asimina* sp.)

Commercial Butterfly Gardens and State Insect Collections

On March 25, 1988, the largest butterfly farm in the world, and the first in North America, opened at Butterfly World in Fort Lauderdale, Broward County. For those interested in viewing not only Florida's beautiful butterflies but also those from around the world, this represents a wonderful opportunity to see the staggering variety of colors and patterns exhibited by butterflies from the U.S. and other countries. Admission is $6 for adults and $4 for children and senior citizens, and opening hours are 9 A.M. to 5 P.M. Monday through Saturday and 1 P.M. through 5 P.M. on Sunday. Butterfly World is located at Tradewinds Park, 3600 W. Sample Road, Coconut Creek, Florida 33073. Although butterfly houses have been popular in Europe since the early 1980s, the Florida butterfly farm is unique in that it can produce many butterflies under their natural tropical conditions without recourse to the hothouses so necessary in cool temperate Europe. Visitors can walk through tropical gardens among thousands of brilliantly colored live butterflies, in all stages of life, within a natural rain-forest environment. At any one time, up to 80 species originally from five continents can be seen. Most will have been actually bred on the site, and visitors can see the breeding facility.

For those in north Florida, another convenient display of living butterflies is provided at Callaway Gardens, Pine Mountain, Georgia, near Columbus. This butterfly house opened on September 25, 1988. It is expected to exhibit 50 species of free-flying tropical butterflies inside the largest glass-enclosed butterfly conservatory in North America. In addition, 65 native species can be found in the beautiful and extensive gardens. Admission is included in the regular Callaway Gardens admission of $4 for adults and $1 for children ages 6–11.

Many collections of pinned butterflies, and other insects, are located in different parts of the state. The Allyn Museum of Entomology, located at 3621 Bay Shore Road, Sarasota, contains a fabulous collection of butterflies from around the world. Lee Miller is chief curator. Access to this collection is limited, however, and preference may be given to "qualified researchers," as is often the case for other collections. Undoubtedly, the biggest collection of Floridian butterflies and other insects is held at the Florida Department of Agriculture and Consumer Services, Division of Plant Industry, at Gainesville. Dr. Harold A. Denmark is chief of entomology there. Tall Timbers Research Station, Route 1 (P. O. Box 678), twenty-five miles north of Tallahassee, also possesses a fine butterfly and moth collection, under the direction of Larry Landers.

Literature Cited

Anonymous. 1921. A case of serious sickness due to the puss moth caterpillar. The Florida Entomologist 4:13–14.

Anonymous. 1945. Nettling caterpillars. The Florida Entomologist 9:45–46.

Arbogast, R. T. 1966. Migration of *Agraulis vanillae* (Lepidoptera, Nymphalidae) in Florida. The Florida Entomologist 49:141–145.

Baggett, H. D. 1982. Threatened Florida atala. Pages 75–77 in R. Franze (ed.), Rare and Endangered Biota of Florida VI. Invertebrates. University Presses of Florida, Gainesville.

Brower, L. P., and J. Z. Brower. 1962. The relative abundance of model and mimic butterflies in natural populations of the *Battus philenor* mimicry complex. Ecology 43:154–158.

Brown, L. N. 1972a. The silkmoths of Florida. The Florida Naturalist 45(2):40–43.

Brown, L. N. 1972b. Stunning Florida moths hatch beneath the earth. The Florida Naturalist 45(4):102–105.

Brown, L. N. 1973. Populations of *Papilio andraeman bonhotei* Sharpe and *Papilio aristodemus ponceanus* Schaus (Papilionidae) in Biscayne National Monument, Florida. Journal of the Lepidopterists' Society 27:136–140.

Buckingham, G. R. 1987. Florida's #1 weed: *Hydrilla* vs. biocontrol. University of Florida, Institute of Food and Agricultural Services, Research 87:22–25.

Buschman, L. L. 1976. Invasion of Florida by the "lovebug" *Plecia nearctica* (Diptera: Biblionidae). The Florida Entomologist 59:191–194.

Byers, C. F. 1930. A contribution to the knowledge of Florida Odonata. University of Florida Publications in Biological Science. Series 1(1):1–327.

Callahan, P. S., and H. A. Denmark. 1973. Attraction of the "lovebug" *Plecia nearctica* (Diptera: Biblionidae) to UV irradiated automobile exhaust fumes. The Florida Entomologist 56:113–118.

Castner, J. L. 1985. Cryptic coloration in the insect world. Florida Wildlife 39(4):9–12.

Castner, J. L. 1986. Delicate balance: Florida atala butterfly, *Eumaeus atala florida*. Florida Wildlife 40(3):39.

Clark, J. T. 1974. Stick and Leaf Insects. Barry Shyrlock and Co., Winchester.

Connor, E. F., S. H. Faeth, and D. Simberloff. 1983. Leaf miners on oak: the role of immigration and *in situ* reproductive recruitment. Ecology 64:191–204.

Correale, S., and R. L. Crocker. 1976. Ground speed of three species of migrating Lepidoptera. The Florida Entomologist 59:424.

Covell, C. V., Jr. 1984. A Field Guide to the Moths of Eastern North America. Houghton Mifflin Company, Boston.

Durin, B. 1980. Insects, etc. An Anthology of Arthropods Featuring a Bounty of Beetles. Hudson Hills Press, New York.

Edwards, G. B., and D. B. Richman. 1977. Flight heights of migrating butterflies. The Florida Entomologist 60:30.

Eisner, T., and S. Nowicki. 1983. Spider web protection through visual advertisement: role of the stabilimentum. Science 219:185–187.

Faeth, S. H., S. Mopper, and D. Simberloff. 1981. Abundances and diversity of leafmining insects on three oak host species: effects of host-plant phenology and nitrogen content of leaves. Oikos 37:238–251.

Fernald, H. T. 1937. An unusual type of butterfly migration. The Florida Entomologist 19:55–57.

Gilbert, I. H., H. K. Gouck, and N. Smith. 1966. Attractiveness of men and women to *Aedes aegypti* and relative protection time obtained with diet. The Florida Entomologist 49:53–66.

Griffiths, J. T. 1952a. Observations on peel injury to Pope summer oranges in the Vero Beach area. The Florida Entomologist 35:127–133.

Griffiths, J. T. 1952b. Some biological notes on katydids in Florida citrus groves. The Florida Entomologist 35:134–138.

Habeck, D. H. 1984. Attracting insects for backyard entomology. The Florida Entomologist 68:117–121.

Hannahs, E. 1986. Butterfly gardening. Florida Game and Fresh Water Fish Commission Bulletin 86/7 NG5.

Harris, L., Jr. 1972. Butterflies of Georgia. University of Oklahoma Press, Norman.

Heidt, J. H. 1954. A report on a case of tick paralysis in Dade County, Florida. The Florida Entomologist 37:149–150.

Hering, E. M. 1951. Biology of the Leaf-miners. Dr. W. Junk, Gravenhages, the Netherlands.

Hetrick, L. A. 1970. Biology of the "love-bug," *Plecia nearctica* (Diptera: Biblionidae). The Florida Entomologist 53:23–26.

Jones, F. M. 1930. The sleeping heliconias of Florida. Natural History 30:635–644.

Jones, C. G., T. A. Hess, D. W. Whitman, P. J. Silk, and M. S. Blum. 1986. Idiosyncratic variation in chemical defenses among individual generalist grasshoppers. Journal of Chemical Ecology 12:749–761.

Kimball, C. P. 1965. The Lepidoptera of Florida. Florida Department of Agriculture, Gainesville. 323 pp.

King, W. V., and L. G. Lennert. 1936. Outbreaks of *Stomoxys calcitrans* L. ("dog flies") along Florida's northwest coast. The Florida Entomologist 19:3–39.

Kingman, S. 1987. Known risk factors explain high rates of infection in Florida. New Scientist 115(1568):20.

Koehler, P. G., and R. S. Patterson. 1987. The Asian roach invasion. Natural History 96:28–35.

Kuitert, L. C., and R. V. Connin. 1952. Biology of the American grasshopper in the southeastern United States. The Florida Entomologist 35:22–33.

Landolt, P. J. 1984. The Florida atala butterfly, *Eumaeus atala florida* Rueber (Lepidoptera: Lycaenidae), in Dade County, Florida. The Florida Entomologist 67:570–571.

Lenczewski, B. 1980. Butterflies of Everglades National Park, Report T-588. National Park Service, South Florida Research Center. Everglades National Park, Homestead.

Leston, D., and V. Waddill. 1980. *Papilio androgeus* (Cramer) (Lepidoptera: Papilionidae) in Dade County, Florida. The Florida Entomologist 63:509.

Lieux, D. B. 1951. Malaria in Florida. The Florida Entomologist 34:131–135.

Lofgren, C. S. 1983. Introduction to the symposium on imported fire ants, southeastern branch, Entomological Society of American, January 26, 1982. The Florida Entomologist 66:92.

Loftus, W. F., and J. A. Kushlan. 1984. Population fluctuations of the Schaus swallowtail (Lepidoptera: Papilionidae) on the islands of Biscayne Bay, Florida, with comments on the Bahamian swallowtail. The Florida Entomologist 17:277–287.

Lubin, Y. D. 1983. *Nephila clavipes* (Araña de Oro, Golden orb-spider). Pages 745–747 in D. H. Janzen (ed.), Costa Rican Natural History. The University of Chicago Press, Chicago.

Madden, A. H. 1945. A brief history of medical entomology in Florida. The Florida Entomologist 28:1–7.

Mallet, J. 1986. Gregarious roosting and home range in *Heliconius* butterflies. National Geographic Research 2:198–215.

May, P. G. 1988. Determinants of foraging profitability in two nectarivorous butterflies. Ecological Entomology 18:171–184.

Meeker, D. 1987. What pests cost Floridians. University of Florida, Institute of Food and Agricultural Services Research 87:11.

Mitchell, P. L. 1980. Combat and territorial defense of *Acanthocephala femorata* (Hemiptera: Coreidae). Annals of the Entomological Society of America 73:404–408.

Morrison, G. 1981. Draft recovery plan for the Schaus swallowtail (*Papilio aristodemus ponceanus* Schaus), with recommendations concerning the Bahamian swallowtail (*Papilio andraemon bonhotei* Sharpe). Final Report to Florida Game and Fresh Water Fish Commission. Tallahassee, Florida. 23 pp.

Moznette, G. F. 1920. Luminous beetles of Florida. The Florida Entomologist 4:17–18.

Muma, M. H. 1967. Scorpions, whip scorpions and wind scorpions of Florida. Florida Department of Agricultures, Gainesville. 28 pp.

Nagano, C., and C. Freese. 1987. A world safe for monarchs. New Scientist 114(1554):43–48.

Needham, J. G., S. W. Frost, and B. H. Tothill. 1928. Leaf Mining Insects. The Williams and Wilkins Company.

Opler, P. A., and G. A. Krizek. 1984. Butterflies East of the Great Plains. The Johns Hopkins University Press, Baltimore.

Patton, C. N. 1956. Observations on the mating behavior of the green june beetle, *Cotinus nitida* (Linn.). The Florida Entomologist 39:95.

Pyle, R. M. 1981. The Audubon Society Field Guide to North American Butterflies. A. Knopf, New York.

Rohani, I. B., and H. L. Cromroy. 1979. Taxonomy and distribution of chiggers (Acarina: Trombiculidae) in north central Florida. The Florida Entomologist 62:363–376.

Scott, J. A. 1972. Biogeography of Antillean butterflies. Biotropica 4:32–45.

Scriber, J. M. 1986. Origins of the regional feeding abilities in the tiger swallowtail butterfly: ecological monophagy and the *Papilio glaucus australis* subspecies in Florida. Oecologia (Berlin) 71:94–103.

Stiling, P. D. 1985. An Introduction to Insect Pests and Their Control. Macmillan Publishers, Ltd.,, Basingstoke.

Stiling, P. D. 1986. Butterflies and other insects of the eastern Caribbean. Macmillan Publishers, Ltd.,, Basingstoke.

Stiling, P. D. 1987*a*. The secret life of the Caribbean. BWIA Sunjet Magazine 11:34.

Stiling, P. D. 1987*b*. Delicate balance: Schaus' swallowtail (*Heraclides aristodemus ponceanus*). Florida Wildlife 41(6):9.

Stiling, P. D. 1988*a*. Eating a thin line. Natural History 97(2):62–67.

Stiling, P. D. 1988*b*. Delicate balance: Florida purplewing (*Eunica tatila tatilista*). Florida Wildlife 42(1):10.

Stiling, P. D., D. Simberloff, and L. C. Anderson. 1987. Non-random distribution patterns of leaf miners on oak trees. Oecologia 73:116–119.

Stirling, F. 1923. Southern migration of butterflies. The Florida Entomologist 7:8–9.

Tissot, A. N., and F. A. Robinson. 1954. Some unusual insect nests. The Florida Entomologist 37:73–92.

Trexler, J. C. 1985. Aggregation and homing in a chrysidid wasp. Oikos 48:133–137.

Urquhart, F. A. 1976. Found at last: the monarch's winter home. National Geographic 150:161–173.

Walker, T. J. 1964. Experimental demonstration of a cat locating orthopteran prey by the prey's calling song. The Florida Entomologist 47:163–165.

Watson, J. R. 1941. Migrations and food preferences of the lubberly locust. The Florida Entomologist 24:40–42.

White, R. E. 1983. A Field Guide to the Beetles of North America. Houghton Mifflin Company, Boston.

Zak, W. 1986. Florida Critters. Taylor Publishing Co., Dallas.

Index

Boldface indicates pages on which illustrations appear, even if the subject of the illustration is dealt with in the text on those same pages.

DATE LOANED